29/-

G. B. Hunter

1217
A2L
Logic

PAST, PRESENT AND FUTURE

PAST, PRESENT AND FUTURE

BY

ARTHUR PRIOR

FELLOW AND TUTOR AT BALLIOL COLLEGE
OXFORD

OXFORD
AT THE CLARENDON PRESS
1967

Oxford University Press, Ely House, London W. 1

GLASGOW NEW YORK TORONTO MELBOURNE WELLINGTON
CAPE TOWN SALISBURY IBADAN NAIROBI LUSAKA ADDIS ABABA
BOMBAY CALCUTTA MADRAS KARACHI LAHORE DACCA
KUALA LUMPUR HONG KONG TOKYO

PRINTED IN GREAT BRITAIN

PREFACE

THIS book is a sequel to *Time and Modality*. Many problems raised in the latter have now been solved, and new ones have been raised in their turn, and I have tried to record some of these developments, and to carry on with some further ones. I have also become aware of the continuing importance of some earlier writings, including some of my own, which I was formerly inclined to think had been simply superseded; so I have something to say about those too. But I have tried to make the book self-contained, presupposing nothing but a few facts, mostly about the better-known systems of modal logic, which can easily be found in the literature.

I have not been able to separate philosophical speculation very sharply from logical computation, and consequently cannot say much that would be helpful to readers who would like to concentrate on the former without too much of the latter—apart from the obvious point that proofs can be skipped without much loss in understanding of what is proved. I would like, though, to make one small but very serious suggestion to readers who are troubled less by symbolism as such than by the particular symbolism employed here: it becomes much more readable if you *don't* all the time try to translate it into some other symbolism, but get into the habit, at least with comparatively short formulae, of reading it directly as English, e.g. read *CFFpFp* straight off as 'If it will be that it will be that *p*, then it will be that *p*', without first turning it into something like $FFp \rightarrow Fp$; and again, read *CGCpqCGpGq* as 'If it will always be that if *p* then *q*, then if it will always be that *p*, it will always be that *q*', without first twisting it into $G(p \rightarrow q) \rightarrow (Gp \rightarrow Gq)$; and read $\Sigma qKNqFKpq$ directly as 'For some *q*, both not-*q* and it will be that both *p* and *q*', instead of trying to get to this *via* something like $(\exists q)\ (\sim q\ \&\ F(p\ \&\ q))$. I have thrown in these direct verbalizations fairly freely throughout the book, and I hope they will be fully used not only as preliminaries to philosophical discussion but also as elucidations of the formulae to which they are attached. But I do not in general give such verbalizations in the middle of proofs, since what is usually important there is just the correct manipulation of the physical shapes, e.g. inserting an L or a G at the beginning of a formula, or before each of the two parts of an implication, by virtue of a given rule.

Most of my indebtednesses are made obvious enough as the work proceeds, but there are some that I should mention here. It was Mr. P. T. Geach who made me aware of the importance of McTaggart, and of the positive aspect of his work; I had thought of him before simply as an enemy. I am grateful to Mr. Geach also, and to Mr. E. J. Lemmon, for a copious correspondence which kept me in touch with new work in the logic of time and of modality when I was back in New Zealand; and in New Zealand itself to Professor J. M. Shorter for his able argumentative presentation of a point of view (discussed in Chapter VII) to which I had been inclined to do less than justice.

A more recent debt is to the University of California in Los Angeles for the opportunity to lecture on these topics there, and to the very lively tense-logicians of California for many discussions with them about their results and mine—notably Nino Cocchiarella in San Francisco, Dana Scott in Stanford, and again E. J. Lemmon, in Claremont. I am grateful also to Ian Hacking and David Berg in Vancouver, to Nicholas Rescher and Storrs McCall in Pittsburgh, to G. H. von Wright, and especially to Charles Hamblin in Sydney, for passing on some of their recent results. And I have learnt much from my students in Los Angeles, particularly Hans Kamp, Patricia Kribs, John Clifford, and Richard Harschman. I suppose that California is the most logically mature place in the world, and now that the logic of tenses is pursued so widely and so vigorously there, its raw pioneering days can be considered over. Actual publications in the field are still, however, small in number, and I hope this book will turn out to be an introduction to a much greater volume of material.

Finally, I would like to thank Dr. A. J. Kenny for many suggestions made after reading the whole book in typescript, and Miss P. Horne and Mrs. M. Heywood for doing most of the typing. I would like to dedicate *Past, Present, and Future* to my colleagues and students in the University of Manchester.

Manchester, 1966 A. N. PRIOR

CONTENTS

I

PRECURSORS OF TENSE-LOGIC

1. *McTaggart's A-series* (*past, present, future*) *and B-series* (*earlier, later*). THE discipline which is now widely called 'tense-logic' is a comparatively new one, and it is worth saying something about its early history while that is recent enough to be accurately remembered. In a sense the founding father of modern tense-logic was J. N. Findlay, who said in a paper published in 1941 that 'our conventions with regard to tenses are so well worked out that we have practically the materials in them for a formal calculus' and that 'the calculus of tenses should have been included in the modern development of modal logics'.[1] But Findlay's remark, like so much that has been written on the subject of time in the present century, was provoked in the first place by McTaggart's famous proof that time is unreal,[2] and we may begin by looking yet again at this celebrated piece of argumentation. For in spite of what seems to me the outrageousness of his conclusion, and the fallaciousness of the reasoning which leads up to it, McTaggart presented what might be broadly called the phenomenology of time with singular accuracy, and drew attention to a body of facts about time which we shall be adverting to frequently in what follows. Indeed, one could say that there is tense-logic itself in McTaggart, though Findlay was the first to see it as such.

'Positions in time', McTaggart says,[3] 'as time appears to us *prima facie*, are distinguished in two ways.' In the first place, 'each position is Earlier than some and Later than some of the others', and 'in the second place, each position is either Past,

[1] J. N. Findlay, 'Time: A Treatment of Some Puzzles', *Australasian Journal of Philosophy*, Dec. 1941, reprinted in A. G. N. Flew's *Logic and Language* (first series, 1951).

[2] This first appeared as an article ('The Unreality of Time') in *Mind* (1908, pp. 457–74), republished in McTaggart's *Philosophical Studies* (London, 1934). It also reappeared in an enlarged form as ch. xxxiii of *The Nature of Existence* (vol. i, Cambridge, 1927).

[3] *The Nature of Existence*, ch. xxxiii, § 305.

Present or Future. The distinctions of the former class are permanent, those of the latter are not. If *M* is ever earlier than *N*, it is always earlier. But an event, which is now present, was future, and will be past.' He then introduces the term 'A series' for 'that series of positions which runs from the far past through the near past to the present, and then from the present through the near future to the far future', and the term 'B series' for 'the series of positions which runs from earlier to later'. He notes that 'the movement of time consists in the fact that later and later terms pass into the present, or—which is the same fact expressed in another way—that presentness passes to later and later terms. If we take it the first way, we are taking the B series as sliding along a fixed A series. If we take it the second way, we are taking the A series as sliding along a fixed B series'.[1]

McTaggart then argues that the B series presupposes the A series, rather than vice versa. His argument starts from the fact that 'time involves change', and that the only way in which events *can* change is in respect of their A-characteristics. If time consisted of a B series only, change could not consist in one event 'ceasing to be an event' while another took its place, for the place of events in the B series is permanent, and so are all their other characteristics and relations except their place in the A series. 'Take any event—the death of Queen Anne, for example—and consider what changes can take place in its characteristics. That it is a death, that it is the death of Anne Stuart, that it has such causes, that it has such effects—every characteristic of this sort never changes. "Before the stars saw one another plain", the event in question was the death of a Queen. At the last moment of time—if time has a last moment —it will still be the death of a Queen. And in every respect but one, it is equally devoid of change. But in one respect it does change. It was once an event in the far future. It became every moment an event in the nearer future. At last it was present. Then it became past, and will always remain past, though every moment it becomes further past.' To this last sentence he adds a comment. 'The past, therefore, is always changing, if the A series is real at all, since at each moment a past event is further in the past than it was before . . . It is worth while to

[1] § 306 and n.

notice this, since most people combine the view that the A series is real with the view that the past cannot change.'[1]

He goes on to consider objections to this argument that might arise from Russell's view of time, according to which 'an assertion that *N* is present' means no more than 'that it is simultaneous with that assertion, an assertion that it is past or future means that it is earlier or later than that assertion. . . . If there were no consciousness, there would be events which were earlier and later than others, but nothing would be in any sense past, present, or future. And if there were events earlier than any consciousness, those events would never be future or present, though they could be past.' As to change, Russell defines this as 'the difference, in respect of truth or falsity, between a proposition concerning an entity and the time *T*, and a proposition concerning the same entity and the time *T*′, provided that these propositions differ only in the fact that *T* occurs in the one where *T*′ occurs in the other'. McTaggart gives the example 'At the time *T* my poker is hot', which may differ as to its truth or falsity from 'At the time *T*′ my poker is hot', and if it does so we may say that there is change.[2]

McTaggart has no difficulty in showing that Russell's translation of propositions about the A series into propositions about the relative positions in the B series of described events and the time of assertion (or of judgement), just will not do. 'The battle of Waterloo is in the past', he points out, is something which was once false and is now true. But 'The battle of Waterloo is earlier than this judgment' is something which is 'either always true, or always false'.[3]

Against Russell's account of change, McTaggart has two arguments of which only one is to the point, whether it is cogent or not, and is as follows: The B series is not the only series of positions 'at' which propositions can be true or false. For example, 'The meridian of Greenwich passes through a series of degrees of latitude. And we can find two points in this series, *S* and *S*′, such that "at *S* the meridian of Greenwich is within the United Kingdom" is true, while the proposition "at *S*′ the meridian of Greenwich is inside the United Kingdom" is false. But no one would say that this gave us change. Why should we say so in the case of the other series?' One might answer,

[1] § 311 and n. [2] § 313. [3] §§ 317–18.

I suppose, that the word 'change' is defined precisely in terms of differences in truth-value between propositions which mention different positions in the B series, and not in terms of differences in truth-value between propositions which mention different positions in any other series. But McTaggart argues that there is nothing so arbitrary about it as this. *These* differences constitute change because they have to do with something first being so and then merely having been so—because the B series is simply a reflexion of the A series.[1] 'Earlier' and 'later' are in fact to be defined in terms of past, present, and future. 'The term *P* is earlier than the term *Q* if it is ever past while *Q* is present, or present while *Q* is future.' This definition, though it is given in a much later chapter of *The Nature of Existence* than the one in which his main argument proceeds,[2] is of some importance here. For it means that anything we want to say in the B-series language can be translated into the A-series language, whereas the converse does not hold (as may be seen from the battle of Waterloo example).

2. *McTaggart's argument against the reality of the A-series.* Having satisfied himself that there can be no time worth the name without an A series, McTaggart goes on to argue that the A series, and therefore time itself, involves a contradiction. The contradiction, as first presented,[3] is simply that (STEP 1) the characteristics of pastness, presentness, and futurity are mutually exclusive, and yet (if the A series is real), 'every event has them all'. This, as it stands, is not very convincing, as McTaggart realizes. 'It is never true, the answer will run, that *M is* present, past and future. It *is* present, *will be* past, and *has been* future. Or it *is* past, and *has been* future and present, or again *is* future, and *will be* present and past. The characteristics are only incompatible when they are simultaneous, and there is no contradiction to this in the fact that each term has all of them successively.'[4] These tensed verbs, however, are said to require explanation, and the explanation, according to McTaggart, is (STEP 2) that 'when we say that *X* has been *Y*, we are asserting *X* to be *Y* at a moment of past time. When we say that *X* will be *Y*, we are asserting *X* to be *Y* at a moment of future time. When we say that *X* is *Y* (in the temporal sense of

[1] § 316. [2] Ch. li, § 610. [3] § 329. [4] § 330.

"is"), we are asserting *X* to be *Y* at a moment of present time.' From the last sentence it is clear that we are to understand the 'being *Y* at' whichever sort of moment it is, as a *non*-temporal 'being'. We must presumably understand similarly the last 'is' in the sentence that follows: 'Thus our first statement about *M*—that it is present, will be past, and has been future—means that *M* is present at a moment of present time, past at some moment of future time, and future at some moment of past time.' But what are a 'moment of present time', a 'moment of past time', and a 'moment of future time'? Pastness, presentness, and futurity cannot *permanently* characterize 'moments' any more than they can permanently characterize events. 'If *M* is present, there is no moment of past time at which it is past. But' (STEP 3) 'the moments of *future* time, in which it *is* past, are equally moments of *past* time, in which it *cannot* be past' (italics mine).[1] So the contradiction is restored in a new guise. 'If we try to avoid this by saying of these moments what had been previously said of *M* itself—that some moment, for example, is future, and will be present and past—then "is" and "will be" have the same meaning as before. Our statement, then, means that the moment in question is future at a present moment, and will be present and past at different moments of future time. This, of course,' McTaggart says, 'is the same difficulty over again. And so on infinitely.'[2]

This seems a perverse conclusion. We are presented, to begin with (in STEP 1), with a statement which is plainly wrong (that every event *is* past, present, and future). This is corrected to something which is plainly right (that every event either *is* future and *will be* present and past, or *has been* future and *is* present and *will be* past, or *has been* future and present and *is* past). This is then expanded (in STEP 2) to something which, in the meaning intended, is wrong. It is then corrected to something a little more complicated which is right. This is then expanded (in STEP 3) to something which is wrong, and we are told that if we correct this in the obvious way, we shall have to expand it to something which is again wrong, and if we are not happy to stop there, or at any similar point, we shall have to go on *ad infinitum*. Even if we are somehow compelled to move forward in this way, we only get contradictions half the time,

[1] § 331. [2] § 332.

and it is not obvious why we should regard these rather than their running mates as the correct stopping-points. But why do we have to make the wrong moves in any case? At least after the first few times, when we've seen the trouble it gets us into, why not pass to the corrected version immediately?

McTaggart's underlying assumption, which generates each of the moves that lead us to a contradiction, appears to be that 'has been', 'will be', and the strictly present-tense 'is' *must* be explicated in terms of a *non*-temporal 'is' attaching either an event or a 'moment' to a 'moment'. McTaggart himself observes, however, that 'propositions which deal with the place of anything in the A series', such as 'The battle of Waterloo is in the past', and 'It is now raining', are of a kind which can be 'sometimes true and sometimes false'. The 'is' that occurs in such propositions therefore cannot be non-temporal. We can perhaps eliminate the oblique tenses by attaching phrases like 'is past' and 'is future' to descriptions of events, so that '*X* has been *Y*' becomes '*X*'s being *Y* is past', '*X* will be *Y*' becomes '*X*'s being *Y* is future', and a more complex example such as '*X* will have been *Y*' becomes 'The being-past of *X*'s being *Y* is future'; but in all these examples the 'being' in 'being *Y*' and in 'being past', and the 'is' in 'is past' and 'is future', must be the present-tense 'being' and 'is' if these expansions are to be accurate. This means that complexes like the being past of *X*'s being *Y*, and the being future of the being past of *X*'s being *Y*, are subject to the same series of mutations as *X*'s being *Y* itself. There is nothing extraordinary or disastrous about this; we do not have to rush to stop it at all costs; it is simply the nature of an A series as McTaggart himself describes it at the beginning of his discussion, and his contradictions arise from trying to turn it into a B series.

One other point should be noticed here. Since the being past, say, of some event, is itself something that can go on in the past, present, or future, and since the being past of something is not a momentary matter but on the contrary, once it has started, is a permanent matter, it is not quite right to say that past, present, and future are 'mutually exclusive' determinations of those things to which they apply. One and the same state of affairs may sometimes obtain in the past, present, *and* future, and is bound to do so if it persists for any length of time. This is true, moreover, not only of such abstract states of affairs as the being past of an event, but also, e.g., of the being hot of a poker. Even in such cases, of course, the being present of the state is one thing and its being past or future another thing, and as regards 'positions' in time (if there are such things) McTaggart's incompatibility thesis no doubt holds.

3. *Broad's criticism of McTaggart; temporal predicates and tenses.* That McTaggart's troubles arise from trying to describe an A-series without using tenses (not even the present-tense 'is') is pointed out in Broad's exhaustive analysis of the argument.[1] Broad goes on to suggest that if we are going to admit one temporal copula ('is'), as it seems we must, we might as well admit the others ('has been' and 'will be'), and drop the temporal predicates 'past', 'present', and 'future'. For the alternatives are (1) to analyse, say, 'It will rain' and 'It has rained' as 'An event characterized by raininess is now future' and '... is now past', which needs one temporal copula and at least two temporal predicates (three if 'It is raining' is to be expanded analogously), or (2) to analyse them as 'An event characterized by raininess will be present' and '... has been present', which needs one temporal predicate and (with the present tense either in the form 'It is raining' or in the form 'An event ... is now present') all three copulas. Nothing is gained by these analyses, and they carry the misleading suggestion that when, say, it has rained, then over and above the raininess which 'has been, and no longer is being, manifested in my neighbourhood', there *is* (non-temporally) a 'rainy event', which 'momentarily possesses the quality of presentness and has now lost it and acquired ... pastness'.[2]

Broad claims even to find a *logical* defect in talk of events, or as he puts it 'event-particles', as 'acquiring presentness' and then losing it. If this did happen, he says, 'the acquisition and loss of presentness by this event-particle is itself an event-particle of the second order, which happens to the first-order event-particle. Therefore every first-order event-particle has a *history* of indefinite length. ... Yet, by definition, the first-order event-particle ... has no duration, and therefore can

[1] C. D. Broad, *Examination of McTaggart's Philosophy*, vol. ii (Cambridge, 1938), ch. xxxv, p. 315.

[2] Ibid., pp. 315–16.

have no history, in the time-series along which presentness is supposed to move.' Broad considers it a merit of J. W. Dunne to have seen that the full development of such a view requires an infinity of higher and higher order time series. He himself blocks this development by distinguishing between genuine 'qualitative change' and what he calls 'absolute becoming'. The phrase 'become present' is only grammatically similar to such phrases as 'become louder', and there is no sense of 'change' which covers both of them. To 'become present', in fact, is 'just to "become" . . . ; i.e. to "come to pass" . . . or, most simply, to "happen" '. Such 'absolute becoming' is presupposed in all change, and therefore cannot be treated as a case of it; probably, indeed, it cannot be analysed at all.[1]

We shall have more to say of this later, but it should be observed here that the problem which Broad sees, if it is one, could arise even *without* dropping into talk of event-particles as 'becoming present', 'becoming past', and so forth. For whatever *is* going on *will have* gone on, and will have gone on longer and longer ago; we are landed with this 'history' of what goes on as soon as we use even such a moderately complicated tense as the future perfect.

4. *Findlay's tense-logical laws.* That part of Findlay's 1941 article which deals with McTaggart charges him, as Broad's examination does, with trying to impose conditions appropriate to a tenseless language upon a tensed one. Findlay insists that there is nothing untidy or illogical about a tensed language as such; on the contrary, even the use of tenses in natural languages is systematic and sure-footed enough to contain (in the words of our first quotation from this essay) 'practically the materials for a formal calculus'. Of the 'calculus of tenses' which he says 'should have been included in the modern development of modal logics', all that Findlay says is that it 'includes such obvious propositions as that

x present = (x present) present
x future = (x future) present = (x present) future;

also such comparatively recondite propositions as that

$(x) \cdot (x \text{ past}) \text{ future}$;

[1] pp. 277–81.

i.e. all events, past, present, and future, *will* be past'. The last law is unfortunately symbolized; the formula suggests that *everything* will have been the case (even permanent falsehoods); but it is easily enough amended to

'((*x* present) or (*x* past) or (*x* future)) → (*x* past) future'.

All of these laws, one suspects, are inspired by McTaggart's discussion. The last reminds us, for example, of McTaggart's initial picture of events which were future becoming present and then moving further and further into the past, and the first two recall the iterations he complains of in 'the Argument'—a future event is one which is future at a present moment and present at a future one; a present event is present at a present moment—only Findlay, instead of complaining of such equivalences and implications and trying to block them (as even Broad does at some points), treats them exactly as they ought to be treated, as laws of the complicated but far from chaotic logic of the A series.

There is a hint of these laws also in the long passage from Augustine's *Confessions* which forms the subject of the earlier part of Findlay's article, though the relevant remarks are in later sections than those on which Findlay concentrates. Since men foresee the future and recall the past, and 'that which is not, cannot be seen', Augustine is tempted to say that even past and future events and moments in some sense 'are', and that there is some 'secret place' from which they come and to which they go. But this, he goes on, will not be of much help after all, for wherever 'time past and time to come' may 'be', 'they are not there as future, or as past, but present. For if there also they be future, they are not yet there; if there also they be past, they are no longer. Wheresoever then is whatsoever is, it is only as present.'[1] '*x* future', in fact, '= (*x* present) future.' The same thing is more directly stated in Aquinas's dictum, commenting on Aristotle, that *praeteritum vel futurum dicitur per respectum ad praesens* ('things are called past and future with respect to the present'), which he explains by adding, *Est enim praeteritum quod fuit praesens, futurum autem quod erit praesens* ('For that is past which was present, and future which will be present').[2] The

[1] Augustine, *Confessions*, bk. xi, chs. xvii, xviii.

[2] Aquinas, *In Aristotelis Libros Peri Hermeneias et Posteriorum Analyticorum Expositio* (Marietti, Turin, 1955), comment on *De Interpretatione*, 16^b 17–19.

dictum is equally well elucidated by the converse point, that what is said to be future is thereby said to be future *now* (and may cease to be so later), and what is said to be past likewise is said to be so *now* (though it may not always have been so); i.e. 'x future = (x future) present' and 'x past = (x past) present'.

5. *Smart's argument that events do not change.* So far as I know, the first attempt to produce a calculus of the type Findlay wanted to see was my own, in the early 1950s. In the intervening ten years, however, much was written which was relevant to such a task, and two items in particular should be noticed. One was J. J. C. Smart's paper 'The River of Time',[1] which was basically hostile to any such enterprise, but helped nevertheless to make clear what had to be done. Smart, like Broad, or at least like Broad in one mood, disliked talk of 'events' as 'changing'. '*Things* change, events *happen*.' Events are indeed said to become present and to become past, but these changes are spurious. That they are so, Smart claimed to show by giving a Russellian analysis of tensed utterances, and showing that this analysis cannot give the same meaning to the tenses of 'to be past' and 'to be future' as it gives to straightforward verbs like 'to be red' and 'to be green'. Saying that (1) a boat '*was* upstream, *is* level, and *will be* downstream', he says, means just 'that occasions on which the boat is upstream are *earlier than* this utterance, that the occasion on which it is level is *simultaneous with* this utterance, and that occasions on which it is downstream are *later than* this utterance'. In this, he observes, 'was', 'is', and 'will be' are correlated with 'earlier than', 'simultaneous with', and 'later than' applied to one and the same utterance. On the other hand, the translation of (2) 'The beginning of the war was future, is present, will be past' is 'The beginning of the war is later than some utterance earlier than this one, is simultaneous with this utterance, and is earlier than some utterance later than this one'. Here the triad of relations is attached to *different* utterances. 'This', he claims, 'shows how misleading it is to think of the pastness, presentness, and futurity of events as properties. . . . It shows how utterly unlike "this event was future and became past" is to "this light was red and became green".'

[1] In *Mind*, Oct. 1949, pp. 483–94, reproduced in Flew's *Essays in Conceptual Analysis* (1956).

This argument, however, is a little sophistical. In the first place, if we literally applied Smart's analysis of example (1) to example (2) we would not get what he says we would, but would get rather: 'Occasions on which the beginning of the war is future are earlier than this utterance, the occasion on which it is present is simultaneous with this utterance, and occasions on which it is past are later than this utterance'; in which the triad of relations *is* attached to the same utterance, exactly as in example (1). Smart only gets his result when he attempts to eliminate not only the three tensed verbs but the adjectives 'future', 'present', and 'past' as well. He has, in fact, equated (2) with (3) 'The war was going to begin, is now beginning, and will have begun', and applied his analysis to the secondary as well as the primary tense-inflexions of these verbs. His equation of (2) with (3) seems to me reasonable enough, and it does suggest that the verbs 'is past', etc., can in general be dispensed with in favour of more complicated tensing of more ordinary verbs. This does not mean, though, that in the more abstract version the simple tenses have to be treated differently from other simple tenses (as shown above, they don't). Nor does it mean that events don't really change; it means only that changes of events with respect to their pastness, etc., are reducible to more complicated changes of less abstract entities with respect to less abstract properties.

Even, however, when we have reduced (2) to (3), it remains true that interior futures and pasts (the 'going to' in 'was going to', and the 'have' in 'will have') do not relate us to the same utterance ('*this* utterance') as the exterior futures and pasts do. But for whom is *this* fact supposed to be awkward? The analysis of the content of tensed utterances in terms of B-series relations to the utterance itself is quite unplausible even when the tenses used are simple, as McTaggart and Broad both saw. But when it is applied to tenses such as the future perfect, it becomes downright fantastic. Where the B-series relation is only supposed to be to the very utterance which is being analysed, the utterance at least in a sense guarantees its own existence, so that it is at least *true* that the event said to be past, say, is earlier than the utterance in question, even if this fact *isn't* (as the theory says it is) what the utterance is intended to convey. But when the analysis requires us to relate the events to *other* utterances, of which there may very well not have been any (or not be going to be any) at the time at which they would be required, it becomes quite obviously wrong. How are we to analyse, for example, 'Eventually all speech will have come to an end'? What Smart's recipe would give is 'The end of all utterances is earlier than some utterance later than this one', which translates something empirically possible into a self-contradiction. It is in any case implausible—as Smart himself insists when presenting this material in the context of his own thesis that events do not change—that the same tenses, used within the same utterance, should take us in one part of the sentence to one utterance and in another to quite a different one. The real moral of Smart's paper is that the Russellian analysis of tenses breaks down, as so many false theories in this area break down, as soon as we remember that there is such a tense as the future perfect.

6. *Reichenbach on the time of speech and the time of reference; the nature of presentness.* Someone who did not forget this, in the late 1940s, was Hans Reichenbach, in the section on 'The Tenses of Verbs' in his *Elements of Symbolic Logic* (1947). Reichenbach learnt from Jespersen that in seeing how tenses work we have to consider not only the time of utterance on the one hand and the time at which the event spoken of occurs on the other, but also a 'point of reference' which may be, though it need not be, different from either. When we say, for example, 'I shall have seen John', the remark directs us, not in the first place to the time at which my seeing of John occurs, but to a time later than that, with reference to which my seeing of John is past. Reichenbach exhibits the characteristic features of this case by the following diagram (where S is the 'point of speech', R the 'point of reference', and E the 'point of event'):

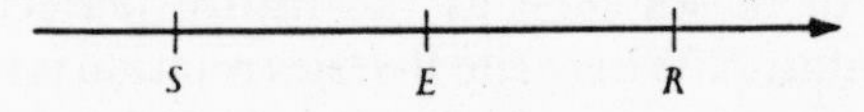

The past perfect, 'I had seen John', comes out analogously as

Jespersen only used this 'three-point structure' to explain these two tenses, but Reichenbach extended it to cover many others, such as the simple past, 'I saw John', which he represents as

and the present perfect, 'I have seen John', which he represents as

This new distinction throws some light on Smart's difficulties with the future perfect, and indeed could be used to construct a partial defence of his point of view. For whereas with the present perfect the pastness expressed by 'have' represents the event's preceding a point of reference which coincides with the point of speech, with the future perfect the pastness expressed by 'have' represents the event's preceding a different point of reference (even if it does not represent its preceding a different utterance). Reichenbach's scheme, however, will not do as it stands; it is at once too simple and too complicated.

It is too simple because, although we do not ordinarily use them, we can easily construct more complicated tenses than the future perfect, e.g. 'I shall have been going to see John'. Here there are in effect two points of reference, which might be (though there are other possibilities) as in the following representation:

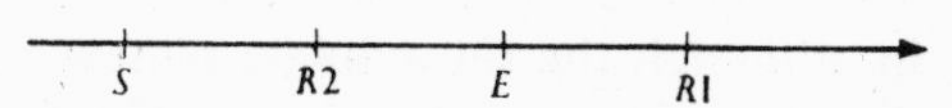

But once this possibility is seen, it becomes unnecessary and misleading to make such a sharp distinction between the point or points of reference and the point of speech; the point of speech is just the *first* point of reference. (This, no doubt, destroys Reichenbach's way of distinguishing the simple past and the present perfect; but that distinction needs more subtle machinery in any case.) This makes pastness and futurity *always* relative to *some* point of reference—maybe the first one (i.e. the point of speech) or maybe some other. Because Reichenbach's analysis fell short of this generalization, it was in some ways a hindrance rather than a help to the construction of a logic of tenses; at all events, no such logic could get going until this generalization had been made. Findlay and his precursors were already ahead of Reichenbach here. His law

'x future = (x future) present = (x present) future', and Aquinas's *Est futurum quod erit praesens*, show a perception that the essence of 'presentness' does *not* lie in coincidence with the point of speech; there is a future presentness and a past presentness also. Broad is closer to the truth (though he draws the wrong moral from what he sees) when he says that to be (or become) present is simply to happen. It is a kind of zero tense-inflexion; the presentness of a happening is simply its happening; Findlay's 'x present = (x present) present' is in fact merely an instance of something still more general, namely 'x = x present' or 'x present = x', from which his law about the future (that the futurity of the presentness of a happening is just the futurity of its happening) also follows.

English speakers find it hard to see these things quite clearly; for in English sentences the point of view of the speaker dominates even subordinate clauses. When an English speaker, for example, wants to say on Tuesday that someone complained on Monday of a sickness that he had that day, the correct form of words will be 'He said he *was* sick', although the man was in fact complaining not of a then-past but of a then-present sickness, and his own words would have been 'I *am* sick'. I am told that in modern Greek it is otherwise, though there is the same change of the pronoun as with us; that is, their wording would be that corresponding to 'He said that he *is* sick'. And indeed in classical Latin, although the subordinate sentence is rendered by an accusative and infinitive, it is the *present* infinitive that is used, *Dixit se esse aegrum* (not *Dixit se fuisse aegrum*). Similarly, on the few occasions on which we use phrases like 'It was the case that', in English, they are not followed by the present but the past; we say 'It was the case that he *was* sick', not 'It was the case that he *is* sick', thus hiding from ourselves the fact that it is the *past presentness* of his being ill, not its past pastness, to which we are alluding. That it is *not* a past pastness is indeed obvious enough to those who know the language; but that it *is* a past *presentness* is perhaps *not* obvious enough, and we are tempted to think that what is now past is perhaps a timeless propositional 'content'.

The *formal* importance of this conception of presentness ('x present = x') is that it underlies, and is required by, the systematic definition of complex tenses in terms of simpler ones.

For suppose we do take the view that tensed utterances can be formed by attaching some sort of modifier to timeless propositional contents, e.g. that 'I shall see John' amounts to something like '(Me seeing John) future', where the element in brackets is supposed to be a non-temporally characterized 'content'. Then if attaching 'future' to such a content forms a future-tense sentence, '(Me seeing John) future' will not itself be the sort of thing 'future' or 'past' can be attached to, since it is not a content but a tensed sentence. The building up of complexes like Findlay's '(*x* past) future' requires that tensing be an operation of which the subjects are themselves tensed sentences, and when we have got inside all other tensing to the 'kernel' of the complex, *its* tense will have to be the present.

These considerations settle immediately the semantic category to which such tense-forming operators must belong. They must be expressions that form sentences from sentences, and so must come out of the same box as the 'not' or 'It is not the case that' of ordinary propositional logic, and the 'Necessarily' or 'It is necessary that' of ordinary modal logic. Findlay had again put his finger on what was needed when he said that a calculus of tenses should have emerged with the 'modern development of modal logics'. In fact, however, it was a new look at an *ancient* development of modal logic which caused the calculus to crystallize.

7. *Time and truth in ancient and medieval logic.* In 1949 P. T. Geach made the following comment in a critical notice[1] of Julius Weinberg's *Nicolaus of Autricourt: A Study in 14th Century Thought*: 'Such expressions as "at time *t*" are out of place in expounding scholastic views of time and motion. For a scholastic, "Socrates is sitting" is a complete proposition, *enuntiabile*, which is sometimes true, sometimes false; *not* an incomplete expression requiring a further phrase like "at time *t*" to make it into an assertion.' Today this has perhaps become a commonplace of logical history, but in 1949 it was quite widely informative. It was certainly informative to myself; I had taken it for granted that it was not only correct but also 'traditional' to think of propositions as incomplete, and not ready for accurate logical treatment, until all time-references had been so

[1] In *Mind*, vol. 58, no. 30 (April 1949), pp. 238–45.

filled in that we had something that was either unalterably true or unalterably false. Geach's remark sent me to the sources. The 'Socrates is sitting' example is not only in the scholastics but in Aristotle, who says that 'statements and opinions' vary in their truth and falsehood with the times at which they are made or held, just as concrete things have different qualities at different times; though the cases are different, because the changes in truth-value of statements and opinions are not properly speaking changes in these statements and opinions themselves, but reflexions of changes in the objects to which they refer (a statement being true when what it says is so, and ceasing to be true when that ceases to be so). This seemed to me to throw a little light on Aristotle's better-known opinion that 'There will be a sea-battle tomorrow' might be (because of the indeterminacy of the situation) 'not yet' definitely true or definitely false. That things might change to being true or false from not being definitely either, is certainly a more radical view than that they might change from being true to being false and vice versa, but it is not as far from this as it is from the view that the passage of time is quite irrelevant to the truth and falsehood of propositions. And in both theories changes in respect of truth and falsehood are thought of as demanded by changes in the fact referred to—from a being so to a not being so (or vice versa) in the simpler case, and from a being indefinite to a being definite in the other.[1]

In 1949 there appeared[2] an article by Benson Mates on 'Diodorean Implication', later incorporated in his book on *Stoic Logic* (1953). This included some material about the views of Diodorus Chronos on the definition of the possible and the necessary. Diodorus seems to have been an ancient Greek W. V. Quine, who regarded the Aristotelian logic of possibility and necessity with some scepticism, but offered nevertheless some 'harmless' senses that might be attached to modal words. The possible, Diodorus suggested, might be defined as what either is or will be true, the necessary as what both is and always will be true, and the impossible as what both is and always will be

[1] Cf. A. N. Prior, 'Three-valued Logic and Future Contingents', *Philosophical Quarterly*, Oct. 1953. I thought then that the logic of tensed propositions could be three-valued and that of tenseless propositions two-valued.

[2] In the *Philosophical Review*, vol. 58 (1949), pp. 234–42.

false (not that these are quite what Quine would offer). He had an argument, to which we shall turn in a later chapter, purporting to show that even on premisses which Aristotelians might be expected to grant, what neither is nor will be true cannot be. Mates, in attempting to formalize the thought of Diodorus, made free use of expresssions like 'p at time t' (Geach, reviewing *Stoic Logic* later,[1] naturally did not miss this, and amplified his remarks on Weinberg); I wondered if it could be done some other way, and tried writing Fp for 'It will be that p', by analogy with the usual modal Mp for 'It could be that p'. Apart from trying to fill in the gaps in the Diodorean 'Master Argument', I was intrigued by another problem. Modern modal logic being full of *dubia* (e.g. does being possibly possible imply being possible?), and presented in the form of a number of alternative systems, one naturally wondered which of these systems the Diodorean definitions would yield. Definitions alone, however, yield nothing at all; to get a logic of the possible from its definition in terms of the future, one must also have a *logic* of futurity. The construction, or at least the adumbration, of a calculus of tenses could not wait much longer.

8. *Symbolism and metaphysics.* The symbolizing of 'It will be that p' in a similar way to the symbolizing of 'It could be that p' and 'It is not the case that p' could in itself have metaphysical, or if you like anti-metaphysical, significance. I did not myself draw much of this out of it until I had done a good deal of 'calculating', but it was there to be drawn. Findlay wrote his 'Time' essay when he was much influenced by Wittgenstein, and Wittgenstein had already said in the *Blue Book* (dictated 1933–4): 'It is the substantive "time" which mystifies us. If we look into the grammar of that word, we shall feel that it is no less astounding that man should have conceived of a deity of time than it would be to conceive of a deity of negation or disjunction.'[2] Nor is that the only substantive that troubles us here by sending us to seek for a corresponding substance. 'Event' is a trouble-maker too, as Broad saw, though he mistook both the trouble and the remedy.

Broad's difficulty about instantaneous 'event-particles' having

[1] In the *Philosophical Review*, vol. 64, no. 1 (Jan. 1955), pp. 143–5.

[2] L. Wittgenstein, *The Blue and Brown Books* (Blackwells, 1958), p. 6.

an indefinitely long history was felt by G. E. Moore also. 'An event which *was* present, *is* past.' And 'every event has, *when* it is present, a characteristic which it does not possess at any other time—a characteristic which is what we mean by saying that at that time and no other it is present'. But against this we may say 'that no event possesses *any* characteristic *at* any time except the time at which it is. . . . It certainly can't be, as language suggests, that the same event *is at* all times, and possesses at one a characteristic which it doesn't possess at others. That would assimilate an event to a thing which persists and has at one time a quality which it hasn't got at others.' The time at which an event 'is present, *means* the time at which it *is*. How can an event have a characteristic at a time at which it isn't?'[1] Broad and Moore are making too much of the *transitory* character of their 'event-particles'; the difference between events and 'persisting things' is more fundamental than that; the real point, one might say, is not that events 'are' only momentarily, but that they don't 'be' at all. 'Is present', 'is past', etc., are only quasi-predicates, and events only quasi-subjects. '*X*'s starting to be *Y* is past' just means 'It has been that *X* is starting to be *Y*', and the subject here is not '*X*'s starting to be *Y*' but *X*. And in 'It will always be that it has been that *X* is starting to be *Y*', the subject is still only *X*; there is just no need at all to think of *another* subject, *X*'s starting to be *Y*, as momentarily doing something called 'being present' and then doing something else called 'being past' for much longer; and no need to argue as to whether *X*'s starting to be *Y* 'is' only at the moment when it does the thing called being present, or also throughout the longer period when it does the other thing. It is *X* which comes to have started to be *Y*, and it is of *X* that it comes to be always the case that it once started to be *Y*; the other entities are superfluous, and we see how to do without them, how to stop treating them as subjects, when we see how to stop treating their temporal qualifications ('past', etc.) as predicates, by rephrasings which replace them with propositional prefixes ('It has been that', etc.) analogous to negation.[2]

[1] *The Commonplace Book of G. E. Moore* (ed. C. Lewy; Allen & Unwin, 1962). Notebook II (*c.* 1926), entry 8 (p. 97).

[2] Cf., with this and with what follows, A. N. Prior, 'Time after Time', *Mind*,

This move also puts to rest the Dunne-ish spectre of an infinity of time-series one within the other. Nothing is left of that one except cases in which one propositional prefix governs another, as in 'It will be the case next year that it was the case 53 years ago that I am being born' (i.e. I will be 53 next year). For this, no special or extraordinary 'will be' (no 'will be' from a new time-series) is required, but just the same old 'will be' that we have in, say, 'It will be the case next year that I am in England'. I can 'be in England' and 'be 53' *at the same time*. (This is the truth, as regards time, behind Newton's 'Times and spaces are, as it were, the places of themselves as of all other things'.) Nor is the interior 'having been', in this example, a special one. There is of course a difference between plain having-been (having been alive for 53 years) and being on the way to having been, just as there is a difference between sitting down and being about to sit down; but the sitting down or the having-been that one is on the way to is just ordinary sitting down or having-been, not a sitting down of some peculiar sort or a having-been in some peculiar time-series. In being 53 next year, i.e. in having then existed for 53 years, what I *shall be* doing is exactly what my older friends *have* done already; not some quite different thing involving a quite different time-series merely because it is governed by a 'shall be'.

Nor do we need still further time-series for recording 'birthdays of birthdays', as when we say 'Next year it will be three years since I was fifty'. Once again we are just piling on prefixes —'It will be the case next year that (it was the case 3 years ago that (it was the case 50 years ago that (I am being born))).' And once again these prefixes are just the ordinary ones. It will be 3 years since I became 50 in exactly the same sense as it will be 3 years since Wilson became Prime Minister; these things happened *in the same year*—not the election in ordinary time and my birthday in super-time—and if we keep our syntax straight, we will find no reason why this should not be so. The formation-rules of the calculus of tenses are not only a prelude to deduction but a stop to metaphysical superstition.

April 1958, pp. 244–6, and *Changes in Events and Changes in Things* (University of Kansas, 1962).

II

THE SEARCH FOR THE DIODOREAN MODAL SYSTEM

1. *The tense-logical basis of Diodorean modal logic.* THE rudimentary tense-logic employed in my own first attempt to analyse the 'Master-argument' of Diodorus[1] was closely geared to the modal systems of von Wright's *Essay in Modal Logic*, which had appeared shortly before (in 1951), though my symbolism was that of Łukasiewicz ($N\alpha$ for 'Not α'; $C\alpha\beta$ for 'If α then β'; $K\alpha\beta$ for 'Both α and β'; $A\alpha\beta$ for 'Either α or β'; $E\alpha\beta$ for 'If and only if α then β'; $M\alpha$ for 'Possibly α'; and $L\alpha$ for 'Necessarily α'). Von Wright subjoined to propositional calculus (with the rules of substitution and detachment) the definition of 'Necessarily α' ($L\alpha$) as 'Not possibly not α' ($NMN\alpha$); the rule of necessitation RL, that if α is any theorem so is Necessarily-α ($\vdash \alpha \longrightarrow \vdash L\alpha$); the modal extensionality rule RE, that if it is a theorem that α is equivalent to β, it is a theorem that 'Possibly α' is equivalent to 'Possibly β' ($\vdash E\alpha\beta \longrightarrow \vdash EM\alpha MB$); and the axioms that if p is true it is possible ($CpMp$), and that possibly either p or q if and only if either possibly p or possibly q ($EMApqAMpMq$). These postulates sufficed for the system he called M; for his system M′, equivalent to the Lewis system S4, he added the axiom that what could be possible is possible ($CMMpMp$); and for his system M″, equivalent to the stronger Lewis system S5, he added the axiom that what could be impossible is impossible ($CMNMpNMp$).

All of von Wright's postulates but the last (the S5 one) are easily seen to be intuitively plausible if we define 'Possibly α' ($M\alpha$) as 'Either it is or it will be the case that α' ($A\alpha F\alpha$). For example, if it is the case that p it either is or will be the case that p ($CpMp$), and if it is or will be the case that it is or will be the case that p, it is or will be the case that p ($CMMpMp$). Moreover, his postulates are not only intuitively plausible but

[1] 'Diodorean Modalities', *Philosophical Quarterly*, July 1955, pp. 205–13.

formally provable, if we adopt a 'logic of futurity' which is exactly similar to his middle system M′ (equivalent to S4), with *F* for his *M*, except for the absence of *CpFp*. This last ('Whatever is, will be') is implausible, but is not needed to obtain *CpMp*, since 'If *p* then either *p* or it will be that *p*' (*CpApFp*) follows from the propositional calculus alone (*CpApq*). Von Wright's S5 axiom *CMNMpNMp* is not only intuitively implausible if read as 'If it is or will be that it neither is nor will be that *p*, then it neither is nor will be that *p*' (i.e. if it is or will be that something will settle into permanent falsehood, it has already done so); it can also be in a manner shown formally to be wrong by deducing from it the even more obviously implausible tense-logical formula *CFNFpNFp* ('If it will be that *p* will never be the case, then *p*—right now—will never be the case').

In order to keep the parallel with von Wright's system exact, a tense-prefix *G*, meaning 'It will always be the case that—', was defined as *NFN* ('It will never be the case that not—'), just as 'Necessarily' (*L*) is defined as 'Not possibly not' (*NMN*). Using this, it was possible to formulate, for example, the rule that if α is a theorem, so is 'It will always be that α'. It is not usual for grammarians to count 'will always' as a special tense, though from a logical point of view it is certainly out of the same box as 'will sooner or later' ('will at some time') which is what the plain 'will' normally means; but whether it be called a tense or not, it is an expression of central importance in the logic of tensed sentences, and originally found its way there through the modal analogy, and from Diodorus, who saw that it had to be used in explicating the 'necessity' that would correspond to his sense of 'possible'.

The handful of postulates so far listed sufficed to show that the Diodorean modal system is at least as strong as Lewis's S4, but does not contain his S5. It was noted, however, that everything in S5 (including *CMNMpNMp*) would be tense-logically plausible if the past as well as the future were brought into the definition of *M*, i.e. if *M*α were read 'α either is or will be *or has been* the case'. (The S5 law *CMNMpNMp* then amounts to 'If it is the case at some time that it is not the case ever that *p*, then indeed it *is* not the case ever that *p*'.) The formal proof of this, however, required a logic of pastness as well as of futurity, and was not attempted in this article.

2. *A matrix for Diodorean modality.* In the 1956 John Locke Lectures on *Time and Modality* (largely prepared in 1955, and published in 1957), the Diodorean concepts of possibility and necessity were represented by an infinite matrix or truth-table, in which 1 and 0 were used to represent truth and falsity at a given instant, and the 'values' assigned to propositional variables were not just 1 and 0 but all infinite sequences of these. The sequence for 'Not *p*' (*Np*) was taken to have 0 wherever that for *p* had 1, and 1 wherever it had 0; and that for '*p* and *q*' (*Kpq*) to have 1 at all points where both the *p*-sequence and the *q*-sequence had 1, and elsewhere 0. The sequence for 'Possibly *p*' (*Mp*) was taken to have 1 at a given point so long as the *p*-sequence had a 1 either there or further to the right, and after that the *Mp* sequence was to have 0's (representing the idea that 'Possibly *p*' is true so long as *p* itself either is or will be so)—for example, if *p*'s sequence is

010001011100 (and then all 0's)

Mp's is

111111111100 (and then all 0's).

The sequence for 'Necessarily *p*' (*Lp*) was to have 1 at a given point if and only if *p* had a 1 from that point on, and elsewhere *Lp* was to have 0 (representing the idea that 'Necessarily *p*' is not true until *p* is and always will be so). For example, with the above *p*-sequence the *Lp*-sequence is 0's all through, and with this for *p*

0100010111001 (and then all 1's)

the *Lp* sequence is

0000000000001 (and then all 1's).

A modal formula was taken to be 'verified' by the matrix if and only if all assignments of sequences to its variables gave the formula as a whole the sequence with 1's throughout (this sequence, that is to say, was 'designated').

This matrix can easily be shown to verify all theses of the Lewis system S4, but not to verify all those of S5. In view of the earlier examination of the Diodorean system, this was to be expected. But the caution of the earlier article was thrown away in *Time and Modality*, and it was asserted there[1] that the system verified by the 'Diodorean' matrix was *precisely* S4, i.e. that the matrix was 'characteristic' for S4, verifying all its theses and no others. At the time this assertion, though unproven, was not quite as rash as it would now appear to be. The only important system that had then been proposed as being possibly weaker than S5 and in any case stronger than S4 was the one which W. T. Parry[2] had called S4.5, which added to S4 the thesis *CLMLpLp*, 'What is necessarily possibly necessary is necessary'. In Diodorean terms this meant that if it is and always will be that it either is or will be that it is and always will be that *p*, then it now is and always will be that *p*. This is complicated, but a little reflection makes it clear that it could be falsified by any *p* which eventually will be true for good, but has not yet quite reached that state. The matrix did falsify S4.5 as well as S5. The assertion in *Time and Modality* was wrong, all the same; and to see one of the points at which it was wrong, a little more should be said about Parry's S4.5.

3. *Modal systems between S4 and S5*. In all the Lewis modal systems, we may use complexes of *L*'s and *M*'s to construct modal assertions of indefinite length, e.g. *LMLLMLMp*. But in S4, owing to such theses as *CLpLLp* and *CMMpMp*, any one of these can be shown to be equivalent to one or other of the following seven, with the implications as shown:

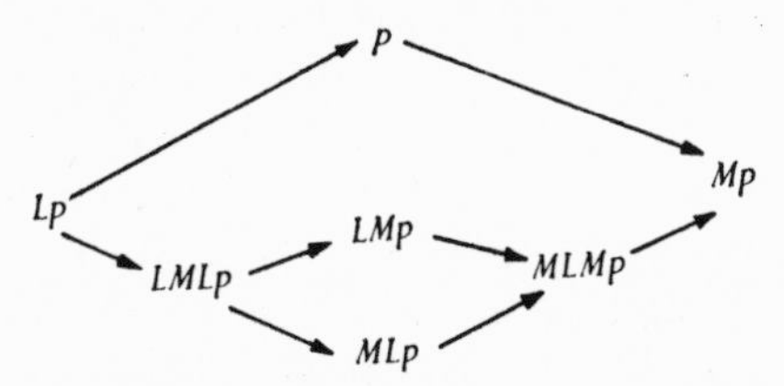

Counting the negations of these as further 'modalities', this gives 14 distinct modalities for S4. But if Parry's S4.5 thesis *CLMLpLp* is added, *LMLp* becomes equivalent (by this and other laws) to *Lp*, and *MLMp* to *Mp*, reducing the number

[1] pp. 23; see also p. 121, n. 1.

[2] W. T. Parry, 'Modalities in the *Survey* System of Strict Implication', *Journal of Symbolic Logic*, vol. 4, no. 4 (Dec. 1949), p. 150.

of distinct modalities at least to 10, namely the following 5 with their negations: *Lp*, *MLp*, *LMp*, *p*, and *Mp*. This would also happen if we collapsed *LMLp*, not upwards to *Lp*, but downwards to *MLp*, and *MLMp* not downwards to *Mp*, but upwards to *LMp*; we would then have the following simple scheme:

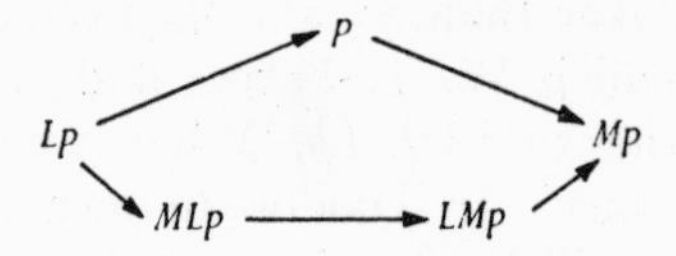

This collapsing will occur if we add to S4 the thesis *CMLpLMLp*, which therefore cannot be in S4, the 14 modalities of that system being known to be not further reducible. I noticed early in 1957, however, that this thesis *CMLpLMLp* is one which the Diodorean conception of modality will verify. This fact is a little clearer with the shorter thesis *CMLpLMp*, from which my one, given S4, is deducible. (This simplification is due to Geach, in 1957.) In Diodorean terms, *CMLpLMp* means that if it is or will be that it is and always will be that *p*, then it is and always will be that it is or will be that *p*. This follows easily from the tense-logical truth that if it will be that it will always be that *p*, then it will always be that it will be that *p*, *CFGpGFp*. The converse of this, it may be observed here, is not the case; *Gp* means 'it will always *uninterruptedly* be' ('it will never not be'), and if *p* is something whose truth and falsehood will always alternate, it will be true to say 'It will always be that it will be that *p*' (*GFp*), but not to say 'It will be that it will always be' (*FGp*), since *p* will never come to be *uninterruptedly* true. Geach called the modal principle *CMLpLMp* the 'quantifier shift' law, because of its structural resemblance to the law of predicate logic that if there is something that everything ϕ's, then everything has something that it ϕ's (though once again not vice versa—'Everyone shaves someone' doesn't imply that there is any one individual that everyone shaves).

Another proof that the Diodorean system is stronger than S4 was discovered a little earlier in 1957 by E. J. Lemmon. His counter-example was the formula *ALCLpLqLCLqLp* ('Either necessarily-*p* necessarily implies necessarily-*q*, or necessarily-*q* necessarily implies necessarily-*p*'). It is not easy either to show intuitively that this formula is Diodorean,[1] or to show that it is not in S4. But it was early found equivalent[2] to another counter-example discovered at the same period by Hintikka, namely *CKMpMqAMKpMqMKqMp*,[3] and although this looks complicated, its tense-logical plausibility is very easy to see indeed. It follows, given the Diodorean definition of *M*, from the tense-logical formula

CKFpFqAAFKpqFKpFqFKqFp,

i.e. if *p* is going to be true (*Fp*), and *q* also (*Fq*), then one or other of three alternatives must obtain: either (1) *p* and *q* will be true together (*FKpq*), or (2) *p* will be true and then *q*, i.e. it will be that (*p*, and it will be that *q*), *FKpFq*, or (3) *q* will be true and then *p* (*FKqFp*). What the Hintikka formula itself says, interpreted in the Diodorean way, is that if it is or will be that *p* (*Mp*) and is or will be that *q* (*Mq*), then either (i) it is or will be that (*p*, and it is or will be that *q*), or (ii) it is or will be that (*q*, and it is or will be that *p*). When the alternatives embedded in (i) and (ii) are fully spread out, they are found to cover precisely the (1), (2), and (3) of the formula in *F*, together with the cases we get when either or each of *p* and *q* is present rather than future.

The Lemmon formula *ALCLpLqLCLqLp* can be shown more easily than the Hintikka one not to be in S4. Lemmon's proof depended on certain relations between S4 and the intuitionist calculus of Heyting which were discovered by Gödel and proved by McKinsey and Tarski.[4] Suppose we 'translate' intuitionist formulae into modal ones as follows: Have all simple propositional variables, and all intuitionist negation and implication signs, immediately preceded by an *L*, and leave conjunction and disjunction signs as they are. For example, such

[1] Such an intuitive proof is given in A. N. Prior's 'Diodorus and Modal Logic: A Correction', in the *Philosophical Quarterly*, July 1958, pp. 226–30. The conjecture at the end of this article is, however, a false one.

[2] A proof of the equivalence is given in A. N. Prior's 'K1, K2 and Related Modal Systems', *Notre Dame Journal of Formal Logic*, vol. 5, no. 4 (Oct. 1964) pp. 299–304. (Strictly, what Hintikka's axiom is here proved equivalent to is Lemmon's as shortened by Geach.)

[3] Hintikka gives a variant of this formula in his review of *Time and Modality* in the *Philosophical Review*, vol. 67 (1958), pp. 401–4.

[4] J. C. C. McKinsey and Alfred Tarski, 'Some Theorems about the Sentential Calculi of Lewis and Heyting', *Journal of Symbolic Logic*, vol. 13, no. 1 (Jan. 1948), pp. 1–15.

a 'translation' of *ANpNNp* would be *ALNLpLNLNLp*, or (since in the modal logic $A = CN$) *CNLNLpLNLNLp*, which the equivalence of *M* and *NLN* contracts to *CMLpLMLp* (my own formula of the last paragraph but one). Again, such a 'translation' of the intuitionist *ACpqCqp* would be *ALCLpLqLCLqLp*—Lemmon's formula. The Gödel–Tarski–McKinsey theorem is that an intuitionist formula is an intuitionist *thesis* if and only if its modal 'translation' is a thesis of S4. In fact neither *ANpNNp* nor *ACpqCqp* are intuitionistic theses, from which it follows that neither my formula nor Lemmon's is an S4 one. They are also—Lemmon's formula as it is, and mine when stated as an alternation—excluded from S4 by a more general consideration. In intuitionistic logic, nothing of the form 'Either α or β' is provable unless either the component α is provable on its own, or β is. From this it follows (given the Gödel–McKinsey–Tarski theorem) that an alternation of modal formulae which 'translates' an intuitionist alternation, will not be in S4 unless one of its alternants is. But neither *LCLpLq* nor *LCLqLp* is a theorem of S4 (we can refute either by putting a logically true formula for the antecedent and a logically false one for the consequent; so not *ALCLpLqLCLqLp*).

The relation of the formulae *ANpNNp* and *ACpqCqp* to the intuitionist calculus was being studied at this time by M. A. E. Dummett. The result of adding the former to Heyting's calculus he called KC, and the result of adding the latter, LC. He was able eventually to show that the full classical propositional calculus contains LC but is not contained in it, and that the same relation holds between LC and KC, and between KC and Heyting's calculus.[1] This result gave an added interest to the modal systems formed by adding *CMLpLMp* (equivalent in S4 to the translation of the KC axiom *ANpNNp*) and *ALCLpLqLCLqLp* (the translation of the LC axiom *ACpqCqp*) respectively to S4. Dummett and Lemmon named the former system S4.2 and the latter S4.3, and showed that they stood between S4 and S5, the latter above the former, exactly as KC and LC stand between Heyting's calculus and classical 2-valued logic.[2]

[1] Michael Dummett, 'A Propositional Calculus with Denumerable Matrix', *Journal of Symbolic Logic*, vol. 24, no. 2 (June 1959), pp. 97–106. (Dummett had these results in 1957.)

[2] M. A. E. Dummett and E. J. Lemmon, 'Modal Logics between S4 and S5',

Some minor results of this period were that, given S4, Lemmon's axiom for S4.3 can be replaced by the slightly shorter *ALCLpqLCLqp* (Geach), and that W. T. Parry's S4.5 is not between S4 and S5 at all but equivalent to S5 (the S5 formula *CMLpLp* is provable in it).

4. *Kripke's 'branching time' matrix for S4, and Lemmon's for S4.2.* In 1958, another contribution was made to the clearing of this jungle. Saul A. Kripke independently communicated a proof that the Diodorean system is not S4. His refuting formulae were *ALMpLMNp* (a variant of the 'quantifier shift' formula) and Hintikka's; and he also gave a matrix which *was* characteristic for S4. For 'values' of the propositional variables, instead of linear series of momentary truth-values, he took *forking* ones or 'trees', and observed that the different branches could be thought of as the different alternative futures that *could* issue from each given point of time. That is, he proposed translating the *Lp* of S4 not as '*p* is true now and will be throughout the *actual* future' but as '*p* is true now and will be throughout all *possible* futures', and its *Mp* not as '*p* either is true now or will be at some point in the actual future' but as '*p* is either true now or will be true at some point in some possible future'. (This, he pointed out, made S4 relevant to the discussion of indeterminism, which was the topic of some later chapters in *Time and Modality*, and at which we shall later be looking again here.) It is easy to see how this model can provide exceptions to Hintikka's S4.3 axiom *CKMpMqAMKpMqMKqMp*. For suppose *p* to be true in some possible future only, and *q* in some other possible future only. We will then have both *Mp* and *Mq* in their two futures, but neither now nor in any possible future do we have *p* either accompanied or followed by *q* (i.e. we do not have *MKpMq*), and neither now nor in any possible future do we have *q* either accompanied or followed by *p* (i.e. we do not have *MKqMp*).

Lemmon has produced a modification of Kripke's model for S4 which distinguishes S4.2 from S4.3. If we use a series of momentary truth-values which indeed may fork, as in Kripke's S4 model, but in which all such divergings are followed by

Zeitschrift für Mathematische Logik und Grundlagen der Mathematik, vol. 5 (1959), pp. 250–64. Lemmon had the results here mentioned by the end of 1957.

later convergings, so that we do have a single line in the end, we can still construct the above counter-example to the Hintikka axiom, but we can no longer construct counter-examples to the S4.2 axiom *CMLpLMp*. We can think of this as representing a time-series in which there are alternative possible *immediate* futures, but only one *ultimate* future. Some theologians, for example, and some Marxists, write as if this is how things are. It should be added, however, that there is a difficulty about the use of Lemmon's S4.2 model to represent this point of view. A time-series that we could diagram as follows:

forks towards the past as well as towards the future, and if there is really only one possible future after the fork, then what that future is, which includes *what will have been the case* in the future, can depend only on 'possible pasts'—one would have to say that once we're past the fork there is no actual past but only the two possible pasts. Some philosophers, indeed, have accepted this consequence. Łukasiewicz, for example, once wrote: 'If, of the future, only that part is real today which is causally determined by the present time; . . . then also, of the past, only that part is real today which is still active today in its effects. Facts whose effects are wholly exhausted, so that even an omniscient mind could not infer them from facts happening today, belong to the realm of possibility. We cannot say of them that they *were* but only that they were *possible*. And this is as well. In the life of each of us there occur grievous times of suffering and even more grievous times of guilt. We should be glad to wipe out these times not only from our memories but from reality. Now we are at liberty to believe that when all the consequences of those fatal times are exhausted, even if this happened only *after* our death, then they too will be erased from the world of reality and pass over to the domain of possibility.'[1] But in general, I suspect, people are much less inclined to talk like this about the past than they

[1] Łukasiewicz, *Z Zagadnien Logiki i Filozofii* (Problems of Logic and Philosophy): *O Determinizmie*, p. 126. My attention has been drawn to this passage, and it has been translated, by P. T. Geach. It is also included in Storrs McCall's forthcoming collection *Polish Logic* (Clarendon Press), pp. 38–39.

are to say that there is no actual future but only various possible futures until we are past the dividing point. But if we don't thus say that *the* past (as opposed to the several possible pasts) is just wiped out at the end of the day, we cannot say that it will *all* be the same in a hundred years' time, no matter what happens in between; since one thing that will be different will be what, by then, *has been* the case. We shall, however, be indicating a mitigation of this conclusion in Chapter VII, Section 5.

5. *Dummett's Formula in D but not in S4.3, and its Presupposition of Discreteness.* To return to Kripke's comments of 1958, he suggested that a correct axiomatization of the Diodorean system, which we may from now on call D, would be obtained by adding the Hintikka axiom to S4. As this axiom expresses the linearity of actual time perspicuously, this looked right. In fact, however, it wasn't, at least if D is taken to be the system for which the *Time and Modality* matrix is characteristic. For Dummett discovered in 1958 a formula which that matrix verified but which could be shown not to be in S4.3 (i.e. the system given by adding Kripke's, Lemmon's, or Hintikka's axiom to S4).[1] This formula was a long one, but was shortened by Geach to

CLCLCpLppCMLpp.

This is still not easy to interpret, but in 1961 I managed to discern the drift of it, how it might be intuitively justified, and why the *Time and Modality* matrix verified it.[2]

By ordinary modal logic the offending formula is equivalent to *CKMLpLCNpMKpMNpp*. For

CLCLCpLppCMLpp	
= *CMLpCLCLCpLppp*	(by *ECpCqrCqCpr*)
= *CMLpCLCNpNLCpLpp*	(by *ELCpqLCNqNp*)
= *CMLpCLCNpMKpNLpp*	(by *ENLCpqMKpNq*)
= *CMLpCLCNpMKpMNpp*	(by *ENLpMNp*)
= *CKMLpLCNpMKpMNpp*	(by *ECpCqrCKpqr*)

Here the component *KpMNp*, if *M*α is defined with Diodorus as *A*α*F*α, is equivalent to *KpFNp*. For in '*p* and it either is or will

[1] Dummett and Lemmon, op. cit., pp. 263–4.

[2] See A. N. Prior, 'Tense Logic and the Continuity of Time', *Studia Logica*, vol. 13 (1962), pp. 133–48.

be that not *p*', the 'is' alternative is not a genuine one, since '*p* and it is the case that not *p*' would be self-contradictory. If the long formula (thus modified) is false, there must be cases in which its two antecedents (*MLp* and *LCNpMKpFNp*) are true, and its consequent (*p*) false. Let us suppose we have such a case, i.e. a *p* for which we have

(1) *MLp*,
(2) *LCNpMKpFNp*, and
(3) *Np*.

Since we have (1) *MLp*, i.e. *ALpFLp*, then either

(1.1) *Lp* already, in which case *p* already; but this is excluded by (3);

or (all that's left)

(1.2) not *Lp* yet, but sooner or later *Lp* (i.e. *p* for ever); therefore sooner or later *p*-false for the last time.

Consider now what happens when we reach the moment when *p* is false for the last time. At this moment we will have *Np*, and therefore by (2) we will have *MKpFNp*, i.e. 'it (is or) will be the case that *p* is true and then false'; so this *isn't* the last moment of *p*'s falsehood. Case (1.2) therefore is as unrealizable as case (1.1), and therefore the combination of (1), (2), and (3) is impossible, and the Dummett formula is a law.

In this argument, however, a dubious step is taken under (1.2). For if time is dense, i.e. if between any distinct moments of time there is an intervening moment, *p* could be false for a while, and then true for ever, *without* there being any last moment of *p*'s falsehood. For there may be a definite first moment of *p*'s permanent truth, and *p* be false *up to* then, in such a way that however close any moment of *p*'s falsehood may be to the first moment of its permanent truth, we can always find a still closer moment at which it is not yet finally true. The Dummett formula is verified by the *Time and Modality* matrix simply because this possibility is not allowed for there, the 'truth-value histories' of propositions being represented in the matrix by *discrete* sequences of momentary truth-values. With this feature discarded, it was possible to re-open the question as to whether S4.3 suffices (as both Kripke and Hintikka had in effect said that it did) for the logic of that 'possibility' which is just presentness-or-futurity, and that 'necessity' which is just presentness-and-permanent-futurity.

That S4.3 does so suffice, and that S4.3 plus the Dummett formula suffice for the system characterized by the discrete matrix, were shown, by different methods, by Kripke in 1963 and by Bull in 1964.[1] The problem of axiomatizing Diodorean modal logic was thereby solved, and in spite of many false moves, a great deal learnt about both time and modality on the way.

We can now for the moment drop modality, and consider what was happening in the meantime to tense-logic itself.

[1] R. A. Bull, 'An Algebraic Study of Diodorean Modal Systems', *Journal of Symbolic Logic*, vol. 30, no. 1 (March 1965), pp. 58–64.

III

THE TOPOLOGY OF TIME

1. *Analysis of the Master-argument of Diodorus.* DIODOREAN modality is defined in terms of the future only, but the Diodorean defence of it, the 'Master Argument', required also some reference to the past. As recorded by ancient writers, the argument is that the following three propositions cannot all be true:

1. Every true proposition concerning the past is necessary.
2. The impossible does not follow from the possible.
3. Something that neither is nor will be is possible.

But the first two are generally admitted; therefore we must deny the third, and admit that whatever neither is nor will be the case is not possible, i.e. that the possible is simply what either is or will be true. To get them into symbolic form, we introduce the following past-tense counterparts of $F\alpha$ and $G\alpha$:

> $P\alpha$ for 'It has been the case that α'
> $H\alpha$ for 'It has always been the case that α'.

The first two propositions in the above allegedly inconsistent triad may be re-worded as follows:

1. Whatever has been the case cannot now not have been the case (*CPpNMNPp*).
2. If p necessarily implies q, then if q is not possible, p is not possible (*CLCpqCNMqNMp*).

And the denial of the third, which is what Diodorus is out to prove from these two, may be represented as follows:

3′. If anything both is not true and will not be true, it is not possible (*CKNpNFpNMp*).

There are clearly some unstated premisses in this proof, and in the second part of my first paper on Diodorus I tried to find reasonably plausible additions which would make the argument valid. Such premisses cannot, of course, include the Diodorean definition of the possible as that which is or will be true; this would give 3′ without any further assistance, but it would hardly be convincing, since it is precisely his rather bizarre definition of the possible that the argument seems designed to defend. And his passage from 1 and 2 to 3′ apparently *was* convincing to the ancients, since the Stoic Chrysippus was driven by it (coupled with his own distaste for 3′) to deny 2, and Cleanthes (who had the same distaste) to deny 1.

The additional premisses which I suggested were the following two:

4. From a thing's being the case it necessarily follows that it has always been going to be the case (*LCpHFp*), or at all events has never-been never-going-to-be the case (*LCpNPNFp*, the preceding expanded by Df. *H*);

and

5. Of whatever is and always will be false (i.e. what neither is nor ever will be true), it has already been the case that it will always be false (*CKNpNFpPNFp*),

for since it is now false and will always be so hereafter, it was the case at least at the moment just gone that it would be always false thereafter. Given these premisses, the argument does indeed take us to the Diodorean conclusion. Schematically, we have

KNpNFp → *PNFp* (by 5)
→ *NMNPNFp* (by 1)
→ *NMp* (by 4 and *CLCpNPNFpCNMNPNFpNMp*, i.e. 2 *q*/*NPNFp*).

Whether the premisses 4 and 5 not only are plausible, and yield the conclusion, but can be found in ancient writers, is a more difficult question to which O. Becker has given some attention since this article appeared.[1] It is quite clear that 4 at least is enunciated and discussed both in Aristotle's *De Interpretatione*, ch. 9, and in Cicero's *De Fato*. Cicero asks, *Potest* . . .

[1] O. Becker, 'Zur Rekonstruktion des "Kyrieuon Logos" des Diodoros Kronos', in *Erkenntnis und Verantwortung* (Festschrift für Theodor Litt, Düsseldorf, 1961), pp. 250–63.

quicquam esse, quod non verum fuerit futurum esse? ('Can that be, of which it was not true that it was going to be?'). Of proposition 5, more will have to be said later.

2. *Some early postulates for past and future.* The logic of the past, and of the past and future together, was treated more systematically than in 'Diodoran Modalities', in an address on 'The Syntax of Time Distinctions' which I read in 1954, and which was published in 1958.[1] There was also, in the later paper, a streamlining of the axiomatization already done, made possible by Sobociński's demonstration in 1953 of the equivalence of von Wright's weakest modal system M to the system T of R. Feys, which takes *L* as undefined, defines *M* as *NLN*, and adds to propositional calculus only the one new rule RL($\vdash\alpha \rightarrow \vdash L\alpha$), and the axioms 'If necessarily *p* then *p*' (*CLpp*) and 'If *p* necessarily implies *q*, then if *p* is necessary so is *q*' (*CLCpqCLpLq*).[2] This suggested re-axiomatizing the system of 'Diodoran Modalities' by taking *G* ('It will always be the case that') as undefined, defining *F* as *NGN* ('It will be true that' = 'It will not always be false that') and adding to propositional calculus the one rule *RG* ($\vdash\alpha \rightarrow \vdash G\alpha$) and the one axiom 'If *p* will always imply *q*, then if *p* will always be the case, so will *q*' (*CGCpqCGpGq*). Sobociński's proofs of T from M were easily adapted to the proof of the earlier tense-logical postulates from these; but it was obvious that neither set was anything like complete for the field.

In the 'logic of futurity' alone, although *CGpp* ('What will always be, already is') and *CpFp* ('What is, will be') are counter-intuitive, their syllogistic product *CGpFp* ('What will always be, will be') seems plausible enough, and so do certain special cases of *CpFp*, namely *CFpFFp* ('If it will be that *p*, it will be—in between—that it will be') and *CNFpFNFp* ('If it will never be that *p*, then it will be that it will never be that *p*'). These are of course the converses of the S4-like thesis *CFFpFp* and the undesirable S5-like one *CFNFpNFp* respectively. In the absence of *CpFp*, the pair *CGpFp* and *CFpFFp* were added to the postu-

[1] In *Franciscan Studies*, 1958, pp. 105–20. ('Diodoran Modalities' appeared in 1955, but was written by early 1954.)

[2] B. Sobociński, 'Note on a Modal System of Feys-von Wright', *Journal of Computing Systems*, July 1953.

lates mentioned above as further axioms, and *CNFpFNFp* proved from them.

To this basis for the 'logic of futurity' a series of analogous postulates were added to give the logic of pastness, together with two special axioms involving both tenses together, namely *CpGPp* ('What is the case will-always have-been the case') and *CpHFp* ('When anything is the case, it has always been the case that it will be the case'). The first of these 'mixing principles' I had found in Ockham's *Tractatus de Praedestinatione*, the reprinting of which by the Franciscan Institute in 1945 had helped to make people aware in that decade of some of the scholastic views on logic and time. Ockham says in this work: *Si haec propositio sit modo vera: Haec res est, quacumque re demonstrata, semper postea erit haec vera: Haec res fuit* ('If this proposition, *This thing is*, be once true, whatever be the object pointed to, then for ever after will this be true: *This thing was*').[1] The other 'mixing principle' *CpHFp* was proposition 4 in my reconstruction of the Master Argument. It is derivable from Ockham's principle, and Ockham's from it, by systematically replacing future-tense symbols (*G* and *F*) by the appropriate past ones (*H* and *P* respectively), and vice versa. It is observed in 'The Syntax' that a rule permitting us to do this with any thesis will cut the axioms by half. Hamblin, using such a rule in 1958, called it a 'mirror-image rule'.

Summing up, the system of this paper adds to propositional calculus the definition of *F* as *NGN* and *P* as *NHN*, the rule RG to infer $\vdash G\alpha$ from $\vdash \alpha$, the mirror-image rule, and the following axioms:

A1. *CGCpqCGpGq* A3. *CFFpFp*
A2. *CGpFp* A4. *CFpFFp*
A5. *CpGPp*

The use of A3 and A4 with *G* undefined is a little inelegant, since these amount by definition to *CNGNNGNpNGNp* and its converse; the equivalent *CGpGGp* and *CGGpGp* would have been better. But these postulates, and slight variations on them, have remained part of the basis of tense-logic in most subsequent formalizations, both my own and other people's. It is now known that they are all independent, and that only one

[1] Franciscan Institute edition, p. 4.

addition is needed to make them complete for an infinite, dense and linear time-series. That they were incomplete was already noted in this paper, it being clear even then that something expressing *linearity* was needed to prove the S5 law *CMNMpNMp* in the tense-logical sense in which it holds, i.e. with $M\alpha$ short for $AA\alpha F\alpha P\alpha$ ('*p* at some time').

Findlay's initial impetus was reflected in the 1954 paper by a proof from the above postulates of his law *CAApPpFpFPp*, 'Whatever is or has been or will be the case, will have been the case'. There is no point in reproducing this proof here, but some deductions may nevertheless be made from the above postulates simply to bring out the structure of the system a little. In the first place, it is clear that RG (to infer $\vdash G\alpha$ from $\vdash\alpha$) and A1 (*CGCpqCGpGq*) will together enable us to pass from any proven implication $\vdash C\alpha\beta$, to the implication $\vdash CG\alpha G\beta$ (we go from $\vdash C\alpha\beta$ to $\vdash GC\alpha\beta$ by RG, and from this to $\vdash CG\alpha G\beta$ by substitution in A1, and detachment). This derived rule (to infer $\vdash CG\alpha G\beta$ from $\vdash C\alpha\beta$) may be called RGC, and can be used to prove others like it. In particular, we have

T1.	*CGCpqGCNqNp*	(*CCpqCNqNp*, RGC)
T2.	*CGCNqNpCGNqGNp*	(A1 *p/Nq*, *q/Np*).
T3.	*CCGNqGNpCNGNpNGNq*	(*CCpqCNqNp*, subst.)
T4.	*CGCpqCNGNpNGNq*	(T1, T2, T3, syll.)
T5.	*CGCpqCFpFq*	(T4, Df. *F*).

This, with RG, gives us the derived rule to infer $\vdash CF\alpha F\beta$ from $\vdash C\alpha\beta$, which we may call RFC. From these results we can further get, by using the mirror-image rule, the rules RHC and RPC, to infer $\vdash CH\alpha H\beta$ and $\vdash CP\alpha P\beta$ from $\vdash C\alpha\beta$. From these (given that $E\alpha\beta$ is just the conjunction of $C\alpha\beta$ and $C\beta\alpha$) we can obtain analogous laws about logical equivalence, i.e. that from $\vdash E\alpha\beta$ we may infer $\vdash EG\alpha G\beta$, $\vdash EF\alpha F\beta$, $\vdash EH\alpha H\beta$, and $\vdash EP\alpha P\beta$ (we could call these rules RGE, RFE, etc.). With all these in our hands, we can carry out such further proofs as the following:

T6.	*EGpGNNp*	(*EpNNp*, RGE)
T7.	*EGNNpNNGNNp*	(*EpNNp*, subst.)
T8.	*EGpNNGNNp*	(T6, T7, E-syll.)
T9.	*EGpNFNp*	(T8, Df. *F*).

T9 means that even though *G* is not defined as *NFN* in this system (but rather *F* as *NGN*), it is still logically equivalent to it. Similarly *H* and *NPN*. We also have

T11. *ENNGNpGNp* (*ENNpp*, subst.)
T12. *ENGpNGNNp* (T6, *CEpqENpNq*)
T13. *ENFpGNp* (T11, Df. *F*)
T14. *ENGpFNp* (T12, Df. *F*).

Similarly, *NP* is logically equivalent to *HN*, and *NH* to *PN*. Further, we have

T15. *EGCpqGNKpNq* (*ECpqNKpNq*, RGE)
T16. *EGNpNFp* (T13, *CEpqEqp*)
T17. *EGNKpNqNFKpNq* (T16, subst.)
T18. *EGCpqNFKpNq* (T15, T16, E-syll.),

i.e. it will always be the case that *p* implies *q*, if and only if it will never be the case that *p* is true together with not-*q*. In none of these proofs, it should be noticed, has any axiom been used but A1 (and its mirror image), and these laws and rules for *G* and *F* on the one hand, and for *H* and *P* on the other, closely parallel laws and rules for *L* and *M* that are analogously provable in the modal system T (= von Wright's M).

Another theorem, using the same very restricted basis, which we will sometimes find useful later, is *CKGpFqFKpq*, 'If *p* will always be true, and *q* will be true sooner or later, then *p*-and-*q* will be true sooner or later', and its image *CKHpPqPKpq*. These correspond to the modal thesis, known to Aristotle, that if *p* is bound to be true and *q* could be, the conjunction of *p* and *q* could be, *CKLpMqMKpq*.[1] We prove it thus:

T19. *CGpCGCpqGq* (A1, Comm)
T20. *CGpCNGqNGCpq* (T19, *CCpCqrCpCNrNq*)
T21. *CGpCNGNqNGCpNq* (T20, *q*/*Nq*)
T22. *ENGCpqFKpNq* (T18, *CEpNqENpq*)
T23. *ENGCpNqFKpq* (T22, *NNp* = *p*)
T24. *CGpCFqFKpq* (T21, Df. *F*, T23)
T25. *CKGpFqFKpq* (T24, p.c.)

[1] Strictly speaking Aristotle uses the allied form *CKMqNMpMKqNp*, i.e. if *q* could be true though *p* is bound to be *false*, we could have the conjunction of *q*-true and *p*-false. (*An. Pr.* 34^a 10–11.)

It should also be noted that A5, *CpGPp*, and its mirror image *CpHFp*, may be replaced by *CFHpp* and *CPGpp*, since

$\vdash CpGPp = \vdash CNpGPNp$ (each from the other by p/Np and if necessary $NNp = p$)
$= \vdash CNGPNpp$ (by $ECNpqCNqp$)
$= \vdash CNGNNPNpp$ (by $NNp = p$)
$= \vdash CFHpp$ (by $F = NGN$, $H = NPN$).
$\vdash CpHFp = \vdash CPGpp$ similarly.

3. *Corresponding postulates in the logic of the B series.* 'The Syntax of Time Distinctions' also contains a systematic correlation of the logic of what McTaggart called the A series (the address's main topic) with that of what he called the B series. In the B-series logic, the propositions of the above system are treated as predicates expressing properties of *dates*, represented by the name-variables *x*, *y*, *z*, etc., *px* being read as '*p* at *x*', and the form *lxy* is introduced for '*x* is a later date than *y*'. An arbitrary date *z* being used to represent the time of utterance, *Fp* is equated with *ΣxKlxzpx* ('For some *x*, *x* is later than *z*, and *p* at *x*', i.e. '*p* at some date later than the date of utterance'), *Pp* with *ΣxKlzxpx* (i.e. '*p* at some date which the date of utterance is later than'), *Gp* as *ΠxClxzpx* ('For all *x*, if *x* is later than *z*, then *p* at *x*', i.e. '*p* at all dates later than the date of utterance') and *Hp* as *ΠxClzxpx*. Given these B-series 'definitions' of the A-series operators, the rule RG, the axioms A1 and A5, and their mirror images, follow by ordinary quantification theory (proof is given, in the paper, for A5 only, but is simple enough for the others). The other postulates are said to be obtainable by putting various conditions on the relation *l*; A3 (*CFFpFp*), in particular, to require the transitivity of *l*; A2 (*CGpFp*), 'the law *Σxlxz*, asserting that there is a date later than any given date'; and A4 (*CFpFFp*), 'the law *ClxzΣyKlxylyz*, asserting that between any two dates there is an intermediate date'. The law *CMNMpNMp* (with *Mα* for *AAαFαPα*) is shown to require the law of trichotomy *AAIxylxylyz*, 'Either the date *x* is identical with the date *y* or it is later than *y* or it is earlier'. Asymmetry, *ClxyNlyx*, is also laid down for 'later than'—reasonably enough—but no law in the tensed system is said or shown to depend on it; and today it seems clear that no law does depend on it.

These are already in principle independence proofs of A3, A2, A4, and *CMNMpNMp*, since each will disappear, and the rest be left, if we remove the condition on *l* which corresponds to it, and the correspondences bring out the special features of time which each axiom expresses—features which in some cases might well be questioned. (Independence proofs for A1 and A5, which express no such special features, are more elusive, but were found by Hacking and Berg in 1965.) Non-transitive temporal succession is perhaps hard to image, but C. L. Hamblin has recently (1965) suggested one such possibility. Suppose time is circular, but as it were changes its sign half-way round. In a cycle taking 3 'aeons', say, to complete, perhaps it will be the case one aeon hence that it will be the case one aeon later that *p*, but that will bring us to a point which is not itself later but one 'aeon' *earlier* than now, so that it is not any interval 'hence', but one aeon *ago*; i.e. although we have *FFp* here, we do not have *Fp* but *Pp*. And *only* A3 (*CFFpFp*) would fail in such a time-scheme.

The correlation of the converse axiom 4, *CFpFFp*, with time's density seems obvious; if time were discrete, then it could be that something will be the case for the last time in the moment that is *just* to come; there will then be no moment at which it will be the case *that* it will be the case (the two 'wills' take us to at least *two* moments hence, and by that time, *ex hypothesi*, *p* is true no more). It should be added, however, that if 'later than' were reflexive, i.e. if every date were later than itself, *lzz*, the law *ClxzΣyKlxylyz* would be trivially verifiable (by putting *z* for our *y*) even if time were *not* dense. It would still imply the B-series version of *CFpFFp*, but now as a special case of *CpFp*, which reflexivity gives very easily. We get reflexivity, of course, if we suppose time to be circular, but adopt, not the convention of the previous paragraph, but the simpler convention that any point we reach by going round in one direction is future (later than now) and any point from which *we* are reached by going round in that same direction is past (earlier than now). Every point than automatically becomes both later than and earlier than itself, and whatever is true will be true (*CpFp*), namely on the next time round. Also, even if time is atomic, 'it will be that *p*' will always imply 'it will be that it will be that *p*' if time is thus circular; for even if *p* ceases to be true just after the next moment, it will start again when we are far enough round.

In correlating A2, *CGpFp*, with time's forward infinity ($\Pi z \Sigma x l x z$), we need to remember that *Gp* is still equivalent to *NFNp*, even if it is no longer that by definition. That is, whether we take *F* or *G* as primitive, the truth-conditions of *Fp* are that it is true if and only if *p* is true at some subsequent moment, and false otherwise, and of *Gp*, that it is false if and only if *p* is false at some subsequent moment, otherwise true. If there is an end of time, then *at* that end, when there are *no* subsequent moments, *Gp* (= 'it will not be the case that not *p*') is vacuously true (*nothing* 'will be' the case then) and *Fp* (= 'it will be that *p*') false. This *G* is in fact like the Boolean version of the Aristotelian form 'Every *X* is a *Y*', which (since it is equivalent to 'Nothing is at once an *X* and not a *Y*') is automatically true if nothing is an *X*. It is easier, all the same, to see that if time has an end we do not always have, and in particular do not have at the end of time, the law *CNFpFNp* ('If it won't be that *p*, it will be that not *p*'); and if the present use of *G* is at this point a little counter-intuitive, the intuitive *Gp* (for which *Gp* as well as *Fp* is false at the end of time, so that *Gp* can even then imply *Fp*) can easily be defined in terms of the present one as *KGpFp* (cf. the definition of a strong 'Every *X* is a *Y*' as the Boolean weak one with 'and something is an *X*').

These considerations apply, *mutatis mutandis*, to the past. If time had a beginning, the mirror image of A2, i.e. *CHpPp*, would have to go; and if time had a beginning but not an end, or vice versa, the mirror-image rule itself would have to go, since we would have one of this pair of mirror images but not the other.

It may be added here that the proof of Findlay's law depends on axiom 2, and that if time has an end, 'It will have been that *p*' is not implied by *p* itself, by 'It has been that *p*', or even by 'It will be that *p*'. For maybe it will be that *p* only at time's last moment, and that is too late for it later to have been that *p*. This is another bit of tense-logic that McTaggart knew about, and summed up in his own way (with its mirror image) by observing that 'if the time-series has a first term, that term will never be future' ('has never been' would have been better), 'and if it has a last term, that term will never be past'.[1]

[1] *The Nature of Existence*, ch. xxxiii, note to § 329.

The proof in 'The Syntax of Time Distinctions' that non-linear time, i.e. time for which we do not have the law of trichotomy *AAIxylxylyx*, would deprive us of *CMNMpNMp*, is a little sketchy and unsure. It is more obvious that non-linearity would deprive us of the law *CKFpFqAAFKpqFKpFqFKqFp* mentioned in the last chapter, and this was used as an axiom to express linearity in some later postulate sets. The 'forking' indeterministic time-series used by Kripke in his model for S4 would be non-linear; the counter-example used in the 1954 address was the time-series of relativity theory. 'Relativity theory distinguishes between an absolute and a relative sense of "later", and if *lxy* means "*x* is absolutely later than *y*", the law of asymmetry holds (no time is at once absolutely earlier and absolutely later than the same time) but the law of trichotomy does not (time *x* may be neither absolutely earlier nor absolutely later than time *y* without being identical with time *y*); whereas if *lxy* only means "*x* is later than *y* from *some* point of view", the reverse is the case.'

These correlations of the 'PF calculus' with an '*l*-calculus' were suggested by the Russellian method of eliminating tenses, but they were not intended to serve the same end, and a caution is given in the paper against treating the arbitrary date *z* of the *l*-calculus as a serious explication of the 'now' which is implicit in the formulae of the other. The interpretation of the latter within the former is, indeed, 'a device of considerable metalogical utility'; and it might have been added that as applied to *theorems* it is harmless, since *l*-theorems are formulae which hold of *any* date *z*, and PF-theorems are formulae which are now and always have been and always will be true. But '"now" is not the name of a date (it has the same meaning whenever it is used, but does not refer to the same date whenever it is used)'. Metaphysically, a translation the other way round would be desirable. 'How this could be achieved in detail has yet to be investigated, but as a first step we may point out that "The date of *p*'s occurrence is later than the date of *q*'s occurrence" seems to be equivalent to "It either is or has been or will be the case that it both is the case that *p* and is not but has been the case that *q*" (*AAKpKNqPqPKpKNqPqFKpKNqPq*).' The negative part of this is perhaps not necessary; without it, the formula is a variation on McTaggart's definition of 'earlier'.

4. *U-calculi and modal logics.* The 'metalogical utility' of associating tense-logical systems with systems developed within predicate logic and the theory of ordering relations is in fact not only 'considerable' but enormous, and something like it (the details vary) is now standard procedure in handling questions of independence and completeness not only in tense-logic, but also, even especially, in modal logic. In some notes made in 1956, C. A. Meredith related modal logic to what he called the 'property calculus' in the following way: Suppose we use a, b, c, etc., as name-variables, and U as a constant 2-place predicate. What the sentence-form Uab means does not matter. It was later suggested by Geach that we might take a, b, c, etc., to name worlds, and Uab to mean that world b is 'accessible' from world a; but again, what 'accessibility' is supposed to mean does not matter. We can treat the sentences of modal logic as if they expressed properties of these objects, i.e. we can use them as predicates in the forms pa, pb, qa, qb, etc. On Geach's interpretation, we can take the specimen form pa to mean that a is a world in which it is true that p. Complex modal sentences express complex properties which are related to complex sentences of the property calculus as follows:

$$(Np)a = N(pa)$$
$$(Cpq)a = C(pa)(qa)$$

(where the N and the C on the left form complex properties, and those on the right form complex propositions).
And

$$(Lp)a = \Pi bCUabpb$$
$$(Mp)a = \Sigma bKUabpb$$

(where Πb means 'for all b' and Σb means 'For some b'); i.e. using Geach's interpretation, 'p is necessarily true in world a' means 'p is true in all worlds accessible from a' (or following the formula more closely, 'For all b, if Uab, pb') and 'p is possibly true in a' means 'p is true in some world accessible from a' ('For some b, Uab and pb'). A modal proposition is a theorem if and only if it is provably true in any arbitrarily chosen world. Different modal systems arise if different conditions are put upon the relation U. If reflexivity alone is imposed, i.e. if our only special axiom for U is Uaa, we obtain von Wright's system

M, or the equivalent system T of Feys. $(CLCpqCLpLq)a$ and the rule to infer $\vdash(L\alpha)a$ from $\vdash(\alpha)a$ follow, given ordinary quantification theory, from the definitions alone (cf. the position of the rule RG and the axiom $CGCpqCGpGq$ in the 'l-calculus'). For the axiom $CLpp$, we expand $(CLpp)a$ first to $C(Lp)apa$, and then to

$$C\Pi bCUabpbpa,$$

which is provable as follows:

C (1)	$\Pi bCUabpb$	
K (2)	$CUaapa$	(1, $C\Pi b\phi b\phi a$)
K (3)	Uaa	(Axiom)
(4)	pa	(2, 3).

(Here we assume the antecedent of the theorem and prove, bit by bit, a conjunction of which the desired consequent is the last member.)[1] If we add further that U is transitive, i.e. if we add the further special axiom $CUabCUbcUac$, we obtain S4. The S4 axiom $CLpLLp$, applied as a predicate to a, gives a proposition which expands first to $C(Lp)a(LLp)a$, and then to $C\Pi bCUabpb\Pi cCUac(Lp)c$ (avoiding unnecessary identifications of variables), and then to

$$C\Pi bCUabpb\Pi cCUac\Pi dCUcdpd.$$

Since consequent-quantifiers binding variables not in the antecedent can be brought to the beginning of an implicational formula, this in turn yields

$$\Pi c\Pi dC\Pi bCUabpbCUacCUcdpd,$$

which is provable as follows:

$\Pi c\Pi dC$ (1)	$\Pi bCUabpb$	
C (2)	Uac	
C (3)	Ucd	
K (4)	Uad	(2, 3, $CUacCUcdUad$)
K (5)	$CUadpd$	(1, $C\Pi b\phi b\phi d$)
(6)	pd	(4, 5).

Make the further addition of symmetry for U, i.e. add also $CUabUba$, and we obtain S5. In Diodorean modal logic, the

[1] For a further account of 'suppositional proofs' of this kind see A. N. Prior, *Formal Logic*, 2nd. ed. (Oxford, 1962), App. II, second part.

'worlds' are clearly instantaneous states of the world, and *Uab* means that *b* is either identical with *a* or one of its temporal successors, and we have to consider what conditions on *U* are appropriate to this interpretation. In my examination in 1961 of the apparently Diodorean Dummett formula

$$CLCLCpLppCMLpp,$$

I found we could only derive this, as applied to an object *a* in a Meredith-style 'property calculus', by assuming for *U* an *inductive* principle which, along with other assumptions, made the time-series appear a discrete one. The principle was constructed as follows: *NUba*, which means that *a* neither is identical with *b* nor succeeds it, amounts to saying that *a precedes b*. And

$$KNUba\Pi cCNUbcUca,$$

i.e. '*a* precedes *b*, and whatever precedes *b* either is identical with *a* or precedes it', amounts (given other assumptions) to saying that *a immediately* precedes *b*. If we abridge this to *Yab*, the inductive principle is

$$C(\Pi bC\phi b\Pi cCYcb\phi c)\ (C\phi a\Pi dCUda\phi d),$$

i.e. if it is the case with every *b* that if ϕ of *b* then ϕ of whatever immediately precedes it, then for any *a*, if ϕ of *a* then ϕ of all its predecessors.[1]

There are connexions between this technique and on the one hand the analogies between modal systems and topological algebras worked out by Tarski and McKinsey, and on the other hand the semantical treatments of modal logic devised by Hintikka, Kanger, and Kripke[2] (some resemblances between the correlations made in 'The Syntax of Time-Distinctions' and his own later work have been drawn out by Kanger);[3] and the methods have perhaps been given their widest generalizations in recent work by Dana Scott and E. J. Lemmon on the 'algebraic' approach to modal semantics.[4]

[1] The proof of the Dummett formula from this and the other assumptions is in 'Tense-Logic and the Continuity of Time' (*Studia Logica*, vol. 13).

[2] See, especially, Kripke's paper on 'Semantical Considerations on Modal Logic', *Acta Philosophica Fennica*, Fasc. 16 (1963), pp. 83–96.

[3] Stig Kanger, review of 'The Syntax of Time-Distinctions', *Journal of Symbolic Logic*, vol. 27 (1962), p. 114.

[4] See, e.g. E. J. Lemmon's 'Algebraic Semantics for Modal Logics', *Journal of Symbolic Logic*, vol. 31 (1966), pp. 46–65 and 191–218; and a forthcoming book, *Intensional Logics*, by E. J. Lemmon and Dana Scott.

Most of these developments are beyond the scope of the present work, but one or two problems which arise may be mentioned. The 'U-calculus' clearly contains many formulae which cannot be put in the form $(\alpha)a$, where α is a formula of tense-logic or of modal logic. The basic conditions which may be put upon the relation *U*, for example—*Uaa*, (reflexivity), *CUabCUbcUac* (transitivity) and so forth—are in general not of this form. And it is not in general necessary that such conditions should entail propositions of this form, i.e. should be 'reflected' in a modal logic or tense-logic, and one might expect some of them not to. It seems in fact that irreflexivity (*NUaa*) and asymmetry (*CUabNUba*), among others, are not so reflected. And there is as yet no systematic way of sorting out conditions on *U* which are thus reflected and ones which are not. It is also often a tricky matter to determine which conditions on *U* are so tied to particular modal or tense-logical theses that the modal or tense-logical system containing them is 'complete' for the type or ordering in question, though the techniques of Scott and Lemmon have greatly facilitated the solution of problems of this sort.

From the point of view of such investigations, a tense-logic is best considered as a species of modal logic with two primitive operators instead of one; though normal tense-logics are not 'modal' in the sense of containing ⊢*COpp* or ⊢*CpOp* for the operators in question. There are, however, weakened modal logics, not containing these theses, which have been studied for their purely formal interest,[1] and some of these may be equated with rather weak tense-logics.

5. *Hamblin's 15-tenses theorem and its basis.* To return now to the consequences that may be drawn from particular tense-logical postulates, an intriguing metatheorem was discovered in 1958 by Charles Hamblin in Sydney. This was a counterpart of the theorem that there are only five non-equivalent affirmative modalities in S4.3; it is to the effect that if we consider any sequence (including the null-sequence) of symbols drawn from *G*, *H*, *F*, and *P* as a 'tense', any possible 'tense' (in this sense)

[1] See, e.g., Ivo Thomas, 'Ten Modal Models', *Journal of Symbolic Logic*, vol. 29, no. 3 (Sept. 1965), pp. 125–8. Thomas's work follows more directly on Meredith's than some of the other items cited.

is equivalent (given certain plausible postulates) to one or other of a group of 15, between which the implication-relations are as follows:

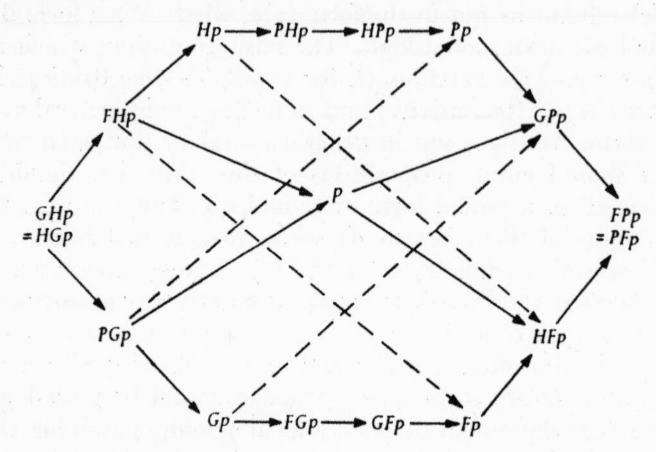

(The dotted implications were not noticed by Hamblin until 1965.) It is interesting to follow these out intuitively. *GHp*, 'It will always be that it has always been that *p*' is clearly true if and only if *p* is omnitemporal, i.e. if it is and always has been and always will be that *p*; and the same is true of its mirror image *HGp*. *GHp* implies *FHp*, since quite generally *G* implies *F*. *FHp*, 'It will be that it has always been that *p*', is true only if it is already the case that *p* has always been true, i.e. *FHp* implies *Hp*, though not vice versa. Similarly, if it is true now that it has always been that *p*, it has been true before that it has always been that *p*, *CHpPHp*, though again not vice versa. These three cases might be diagrammed as follows, with the vertical line representing the present moment, and the covering strip the times at which *p* is said to be true:

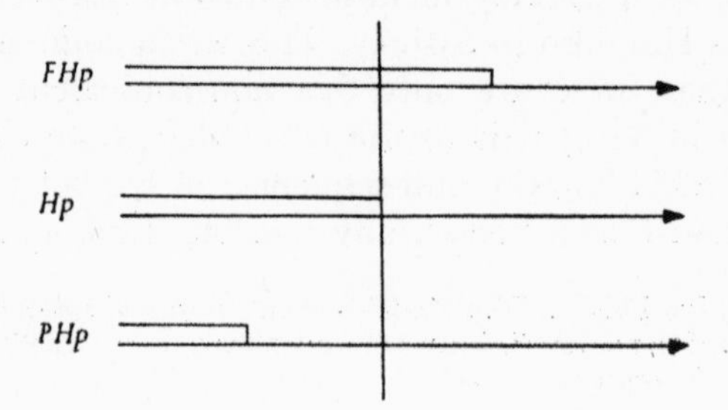

If it has been that it has always been, PHp, then it has always been that it has been, HPp, though not vice versa; HPp might be true and PHp false if p has always been true *on and off*, but never uninterruptedly (cf. the discussion of MLp and LMp in the preceding chapter). If it has always been that p has been, then p has been, $CHPpPp$. If p has been so (and also if it is), it will-always have-been-so, $CPpGPp$ and $CpGPp$ (cf. Ockham). And finally, what will-always have been so will have been so, $CGPpFPp$. (FPp, as in Findlay's law, is the 'deposit' left by all the rest; it is true if we find p true at any time at all). The lower 8 are of course the mirror images of the upper ones. Also, the right-hand 8 are 'duals' of the left-hand 8, i.e. if we have ϕp on the left, we have an equivalent of $N\phi Np$ at the reflecting position on the right; e.g. $HPp = NPHNp$ (for $HP = NPNNHN = NPHN$), and $GPp = NFHNp$ ($Gp = NFNNHN = NFHN$). If we wish to prove that in every case the prefixing of a new symbol will yield an equivalent of something already in the table, we only need consider one quadrant, since the corresponding equivalences with mirror images and duals easily follow. With the top left-hand quadrant the effect of such prefixing works out as follows:

	HG	*PG*	*G*	*FG*
P	*HG*	*PG*	*PG*	*FG*
H	*HG*	*HG*	*HG*	*FG*
F	*HG*	*FG*	*FG*	*FG*
G	*HG*	*G*	*G*	*FG*

i.e. $PHG = HG$, $PPG = PG$, $PG = PG$, etc. The first column is easy; it follows from the fact that if p is true at all times, it is true at any time *that* it is true at all times. Some of the rest are easy, some more difficult, to establish; but in all cases the proofs of the equivalences must rest, in the end, on rules and axioms.

The basis used by Hamblin was adjusted to the use of F and P as primitives, with G and H defined as NFN and NPN respectively. He subjoined to the propositional calculus three rules—to infer $\vdash NFN\alpha$ from $\vdash\alpha$, to infer $\vdash EF\alpha F\beta$ from $\vdash E\alpha\beta$, and the mirror-image rule; and for axioms, the implication

H1. *CNFNpFp* (*CGpFp*) and four equivalences which he stated semi-verbally as follows:

H2. *F*(*p* or *q*) = *Fp* or *Fq* H4. (*p* or *Pp*) = *NFNPp*
H3. *FFp* = *Fp* H5. (*p* or *Pp* or *Fp*) = *FPp*.

H5, *EAApPpFpFPp*, is Findlay's law combined with its converse *CFPpAApPpFp*; this latter could have replaced H5, and is a particularly powerful formula; it constitutes one way of reflecting time's linearity. With this 'von Wright' type of axiomatization, however, it is easier to proceed from equivalences than from implications. For example, we have the following proof (retaining Hamblin's 'equational' presentation, but condensing the equated items):

1. *AANpPNpFNp* = *FPNp* (H5 *p*/*Np*)
2 *NAANpPNpFNp* = *NFPNp* (1, *CEpqENpNq*)
3. *KKNNpNPNpNFNp* = *NFPNp* (2, *NApq* = *KNpNq*)
4. *KKpHpGp* = *NFPNp* (3, Df. *H*, Df. *G*, *NNp* = *p*)
5. *PNp* = *NNPNp* (p.c.)
6. *FPNp* = *FNNPNp* (5, second rule)
7. *NFPNp* = *NFNNPNp* (6, *CEpqENpNq*)
8. *NFPNp* = *GHp* (7, Df. *G*, Df. *H*)
9. *KKpHpGp* = *GHp* (4, 8, *E*-Syll.)

This also combines an easily provable implication, this time *CGHpKKpHpGp*, with a very powerful one, *CKKpHpGpGHp*, which is useful as an axiom in systems with *G* and *H* as primitives.

H4, *EApPpGPp*, can be split into the two implications *CGPpApPp* and *CApPpGPp*, and the latter split further into *CpGPp* (Ockham) and *CPpGPp*. The last of these is superfluous, since substitution in the preceding yields *CPpGPPp*, which can be condensed to *CPpGPp* by a suitable use of the mirror image of H3 (*EPPpPp*). The remaining component, *CGPpApPp*, is interesting. It tells us that not only will it be 'always true that it has been that *p*' if *p* either is or has been true, but if it will always be true that *p* has been true, then *p* either is true or has been. Contraposing this, if *p* neither is nor has been true, it will not always be true that *p* has been true; and giving its mirror image, if *p* neither is nor will be true, it has not always been true that *p* will be true. The force of this last (*CKNpNFpNHFp*) may be given by slightly rephrasing it thus: If *p* neither is nor ever will be true, then it hasn't been true right up to the last moment that *p* will be true. And that's the plausibility of it—if *p* is now and always will be false then it has already been true in the past, at least at the moment *just* past, that *p* will never be true any more—it hasn't always been true, because at least in the moment just past it wasn't true, that *p* would ever be true again.

This is precisely Proposition 5 in the reconstruction of the Master Argument of Diodorus. And it is interesting to be given a basis for tense-logic from which it is provable. Just this Proposition 5, however, had begun about 1960 to strike me as dubious. Theses which appeal, in order to gain intuitive plausibility, to what was the case at 'the moment just past', are liable to commit one to the view that time is discrete. What if there is *no* 'moment just past', but between any past moment, however close to the present, and the present itself, there is another moment still past? On this supposition, Proposition 5 in fact fails, and on the corresponding supposition about the future, Hamblin's H4 fails too. It could be that *p* is now false for the first time, though it will never be true again; and in this case it *has* always been true that *p* will be true; even in the very near past, bringing us as close as we like to the first moment of its falsehood, 'it will be true' must still have a tiny interval to verify it. As to H4, 'It will always be that *p* has been true' certainly does imply that '*p* has been true' will be true in the very near future; but however near we make it, this is still compatible with *p*'s being false now and throughout the past, i.e. with it being false that *p* either is or has been true.

What is most awkward about Hamblin's basis, however, is not this thesis in itself, with its suggestion that time is discrete, but its combination with H3, *EFFpFp*, which as strongly suggests that time is *not* discrete. The system is not actually inconsistent; as T. J. Smiley pointed out, its postulates all come out true if we let $Fp = Pp = p$ (instantaneous time), and we shall see that they also do so on a less radical re-interpretation. But it is not a very happy intuitive basis for proving the theorem of the 15 tenses. Fortunately it turns out that the theorem can be equally well proved if the equivalence H4 is replaced by the corresponding one-way implication *CApPpGPp*, or (to cut out superfluities) simply by *CpGPp*. It *cannot* be proved, however, if H3, *EFFpFp*, is dropped or weakened. If time is discrete, and *Fp* does not entail *FFp*, there are not 15 but an indefinite number of distinct tenses, even using *F* alone. For *Fp* would be true, and *FFp* false, if *p* were going to be true at the next moment for the last time; *FFp* true and *FFFp* false if *p* were going to be true for the last time at the next moment but one; and so on.

6. *Cocchiarella's tense-logic, and differences between linear and branching time.* Another basis for tense-logic with *P* and *F* as primitives was provided in 1965 by N. B. Cocchiarella.[1] Not wishing to commit *logic* to either the discreteness *or* the denseness of time, Cocchiarella dropped the axiom *CFpFFp*; not wishing to commit it to time's being infinite both ways any more than to its not being so, he also dropped *CGpFp*; and not wishing to commit it to time's being altogether similar (e.g. in respect of infinity) in both directions, he dropped the mirror-image rule and simply gave the mirror-images of his axioms separately. This left him, as far as purely propositional tense-logic went (he also had postulates for tensed predicate-logic and identity theory), with the rules to infer $\vdash NPN\alpha$ and to infer $\vdash NFN\alpha$ from $\vdash\alpha$, and the axioms

C1.1. *CNPNCpqCPpPq* C1.2. *CNFNCpqCFpFq*
C2.1. *CpNFNPp* C2.2. *CpNPNFp*
C3.1. *CPPpPp* C3.2. *CFFpFp*
C4.1. *CKPpPqAAPKpqPKpPqPKqPp*
C4.2. *CKFpFqAAFKpqFKpFqFKqFp*
C5.1. *CFKpPqAAFKqFpKqFpKPqFp*
C5.2. *CPKpFqAAPKqPpKqPpKFqPp*

Here the C1's are the appropriate variants of *CGCpqCGpGq*, and the C2's of *CpGPp*; the C3's are familiar expressions of the transitivity of 'earlier' and 'later', and the C4's of time's linearity. The C5's are like the last, but bring both tenses in; C5.1, for example, says that if it has been the case that (*p* is true and *q* will be), then either (1) it has been the case that (*q* is true

[1] References to this writer are to his Ph.D. thesis for the University of California in Los Angeles, entitled 'Tense and Modal Logic: A study in the topology of temporal reference'.

and p has been), or (2) q is true now and p has been, or (3) q will be true and p has been.

The conception of logical 'purity' underlying Cocchiarella's excisions is a questionable one. It has indeed been sometimes said that tense-logic is really not logic but physics, or that it has a good deal of physics 'built into it'. Perhaps it is; the line between logic and other subjects seems to me in any case not an easy one to draw except arbitrarily, and it's not difficult to think of arbitrary ways of drawing it that would exclude the operators P and F altogether (and these would be not *very* arbitrary ways at that). But it seems a bad way to draw the line if there are admitted truths expressible solely in certain terms (e.g. with no constants but P, F, and truth-functions) on both sides of it, so that some truths expressible in one and the same technical vocabulary count as 'logical' truths and some do not. Perhaps, of course, the physics (if that's what it is) is bad physics; and the truths (if that's what they are) not admitted very widely, or not by the experts; and that's more serious. But if we want to be really safe, it's odd to begin by insisting on linearity, and it might be better (as Lemmon has suggested) to confine one's 'basic' laws to those which put no special assumptions on the earlier-later relation at all, i.e. the rules and the C1's and C2's (though even these we shall later find reasons to query). Lemmon calls this 'minimum' system K_t.

Cocchiarella's limited system, whatever the justification for the limitations, has its features of interest. It is too weak to prove the 15-tenses theorem, but it is strong enough for its Diodorean-modal fragment (i.e. the logic of M and L with $Mp = ApFp$ and $Lp = KpGp$) to be still S4.3. It is also strong enough for its modal fragment with $Mp = AApFpPp$ and $Lp = KKpGpHp$ still to be S5. Infinity and denseness, in other words, i.e. $CGpFp$ and $CFpFFp$ and their images, do not yield any special modal theorems, in either of the tense-logical senses of modality. And all the assumptions used by Cocchiarella are needed—you won't get the results just mentioned in anything weaker.

Not all the *axioms* are needed however. Later in 1965 I was able to show that the C5's could be replaced by a shorter pair—C5.1 by $CKKpHpGpGHp$, and C5.2 by its mirror image $CKKpGpHpHGp$. Or alternatively (and more neatly when P and

F are the primitives), by the equivalent transposed forms *CFPpAApPpFp* and *CPFpAApPpFp*. The proof of C5.2 from the latter is as follows:

1.	*CPFpAApPpFp*	
2.	*CKpFqKFqGPp*	(*CpGPp*)
3.	*CKFqGPpFKqPp*	(*CKFpGqFKpq*)
4.	*CKpFqFKqPp*	(2, 3, Syll)
5.	*CPKpFqPFKqPp*	(4, RPC)
6.	*CPKpFqAAKqPpPKqPpFKqPp*	(5, 1)
7.	*CFKqPpFq*	(p.c., RFC)
8.	*CPKpFqPp*	(p.c., RPC)
9.	*CPKpFqCFKqPpKFqPp*	(7, 8, p.c.)
10.	*CPKpFqAAKqPpPKqPpKFqPp*	(6, 9, p.c.).

(10 is just C5.2 with the first two alternants exchanged.) Lemmon then showed that my abridgements were in turn derivable from C4.1 and C4.2, so that the C5's are simply superfluous. (This leaves the C2's as the only 'mixed' axioms, and in fact the remaining ones in *F* are complete for linear and transitive futurity, and those in *P* for the like in pastness—another Lemmon result.) Lemmon's proof of *CKKpGpHpHGp* from C4.2 is as follows (using *Lp* for *KKpGpHp*):

1.	*CKFNpFLpAAFKNpLpFKNpFLpFKLpFNp*	(C4.2)
2.	*CKFNpFLpFAAKNpLpKNpFLpKLpFNp*	(1, *CAFpFqFApq*)
3.	*CPKFNpFLpPFAAKNpLpKNpFLpKLpFNp*	(2, RPC)
4.	*CKLpPFNpKHFLpPFNp*	(*CpHFp*)
5.	*CKLpPFNpPKFNpFLp*	(4, *CKHpPqPKpq*)
6.	*CKLpPFNpPFAAKNpLpKNpFLpKLpFNp*	(5, 3)
7.	*CKNpFLpKNpFHp*	(*CLpHp*, from Df. *L*; RFC)
8.	*CKNpFLpKNpp*	(7, *CFHpp*)
9.	*NKNpFLp*	(8, *NKNpp*)
10.	*NKLpFNp*	(*CLpGp*, from Df. *L*)
11.	*NKNpLp*	(*CLpp*)
12.	*NAAKNpLpKNpFLpKLpFNp*	(11, 9, 10)
13.	*HGNAAKNpLpKNpFLpKLpFNp*	(12, RG, RH)

14. *NPFAAKNpLpKNpFLpKLpFNp* (13, *HGN* = *NPF*)
15. *NKLpPFNp* (14, 6)
16. *CLpNPFNp* (15)
17. *CLpHGp* (16)

It is interesting to group the axioms which go together in these deductions. They are:

I { C4.1. *CKPpPqAAPKpqPKpPqPKqPp*
CKKpHpGpGHp
C5.1. *CFKpPqAAFKqFpKqFpKPqFp* }

II { C4.2. *CKFpFqAAFKpqFKpFqFKqFP*
CKKpGpHpHGp
C5.2. *CPKpFqAAPKqPpKqPpKFqPp* }

If the true picture of time is given by the branching futures of Kripke's model for S4, so that there are alternative routes into the future but only one way back from any point into the past, all the laws of Group I remain, but all those of Group II cease to hold. As we saw earlier, if we find *p* and *q* in separate possible futures and nowhere else, it both 'will' be that *p* and 'will' be that *q*, but in no possible future do we have either *p* and *q* simultaneously, or *p* and then *q*, or *q* and then *p* (refuting C4.2). Again, suppose that *p* is true now, and always has been true but has in the past had chances of being false which it just has not taken; it has no more of these now, however (it has got set in its ways, 'addicted' to being true), and will be true throughout all possible futures (thus verifying *Gp* as well as *Hp* and *p*). Under these conditions *GHp* will be true—throughout all possible futures now, one will on looking back have to say that *p* has always been true—and hence *CKKpHpGpGHp* will be verified. But *HGp* will *not* be true—it has *not* always been the case that *p* would be true throughout all possible futures; there were once possible futures in which it was going to be false. Hence *CKKpGpHpHGp* is here no law. Finally, for C5.2, suppose that it was once the case that *p* was true and also that *q* was going to be true in one possible future; that possibility, however, has not materialized, and we find neither of them true anywhere else in the picture. Then '*p* is true and *q* "will" be' (i.e. in some possible future) was once the case (*PKpFq*); but it is not true either that it has been the case that *q*, and *p* before that (*PKqPp*), or that *q* is true now and *p* has been (*KqPp*), or that *q* 'will' be true and *p* has been (for *q won't* be now, even in a merely possible future).

Cocchiarella himself considers a time-series in which there are divergent paths in both directions, so that we have neither the formulae of Group II nor those of Group I. He identifies this with the 'causal time' of relativistic physics, and notes that if the Diodorean necessity is defined in terms of *this* time-series its postulates are not those of S4.3 but those of S4.[1] It might be added, further, that with this non-linear time the definition of $L\alpha$ as $KK\alpha G\alpha H\alpha$ does not yield S5, or even S4—with this definition, a linearity axiom is required even for the proof of *CLpLLp*. On the other hand, even Lemmon's minimal system K_t (and *a fortiori* Cocchiarella's 'causal' system, which does have *CGpGGp*) yields with this definition of *L* the formula *CpLMp*, which is not in S4. The proof is as follows:

C (1)	*p*	
K (2)	*AApPpFp*	(1, *CpApq*)
K (3)	*GPp*	(1, *CpGPp* from K_t)
K (4)	*GAApPpFp*	(3; *CqApq*, *CpApq*, RGC)
K (5)	*HFp*	(1, *CpHFp*)
K (6)	*HAApPpFp*	(5; *CqApq*, RHC)
K (7)	*KK*(2)(4)(6)	
(8)	*LMp*	(7, Df. *M*, Df. *L*).

CpLMp is a characteristic thesis of what is sometimes called the 'Brouwersche' modal system, which is in between T and S5, and independent of S4. It has been studied by Hintikka and Kripke, and corresponds to a *U*-logic in which the sole conditions on *U* are reflexiveness and symmetry. With $L\alpha$ for $KK\alpha G\alpha H\alpha$, the *L*-fragment of the system K_t is exactly the Brouwersche system (Lemmon, 1966). Whether, with this *L*, the *L*-fragment of Cocchiarella's system is also the Brouwersche system or something between it and S5, is not known; but certainly the Diodorean definition of $L\alpha$ as $K\alpha G\alpha$ gives different

[1] Cf., J. Hintikka, in 'The Modes of Modality', *Acta Philosophica Fennica*, Fasc. 16 (1963), p. 76. Cocchiarella discusses the point, not in the final version of his thesis, but in an abstract, 'Modality within Tense Logic', forthcoming in the *Journal of Symbolic Logic*.

L-fragments with the two tense-logics—with K_t, not S4, but only the system T (Lemmon).

That $L\alpha = KK\alpha G\alpha H\alpha$ does not give S5 in non-linear time was already noted in 'The Syntax of Time Distinctions'; that it does not even give S4 is a new result, but there is a closely allied result in Carnap's development of what he calls 'space-time topology'.[1] Carnap here gives axioms for the earlier-later relation in the 'local proper time' of relativistic physics, laying it down that this relation is transitive, irreflexive, dense, infinite both ways, and not branching in either direction. He then defines 'genidentity' as the logical sum of identity, earlier-than and later-than (i.e. *x* and *y* are genidentical point-instants if either $x = y$ or *x* is earlier than *y* or *x* is later than *y*); and he is only able to prove that genidentity is transitive (the property corresponding to the assertion of *CLpLLp* for the above-defined *L*) by using the axioms which exclude branching. On the other hand, to prove that genidentity is symmetrical (the property corresponding to the assertion of the 'Brouwersche' thesis *CpLMp*) and reflexive (the property corresponding to *CLpp*) he only has to appeal to the definition of this relation.

7. *Further simplifications by Scott and Lemmon.* A slightly less cautious tense-logic than Cocchiarella's, but otherwise more compact, has been presented by Dana Scott.[2] Scott takes *G* and *H* as primitive (with *F* defined as *NGN* and *P* as *NHN*), and (subjoining as usual to the propositional calculus with substitution and detachment) has the rules to infer $\vdash G\alpha$ and $\vdash H\alpha$ from $\vdash\alpha$, and the axioms *CGCpqCGpGq*, *CpGPp*, *CGpFp*, *CGpGGp*, and *CKKpGpHpHGp*, with their mirror images. The system is thus committed to time's infinity (by *CGpFp*) but not to its density (lacking *CGGpGp*). Its most interesting feature is the representation of linearity by the comparatively short *CKKpGpHpHGp* and its mate; Scott having been able to prove the longer Hintikka-style axioms from these. One gets a system equivalent to Hamblin's (corrected) by adding *CGGpGp*.

Certain further economies are possible at another point.

[1] R. Carnap, *Introduction to Mathematical Logic* (1954; translation 1958), Part II, ch. 9. The main results of this chapter are already in his *Abriss der Logistik* (1929).

[2] In a Hume Society talk (with thermofaxed summary, Stanford University) entitled 'The Logic of Tenses' (Dec. 1965).

If we drop the mirror-image rule, it is not necessary to lay down the mirror images of *all* the axioms. In particular, if we have Lemmon's K_t complete (i.e. rules RG and RH; Cocchiarella's C1's and C2's, or my own A1 and A5 with their images), and *CGpGGp* (= *CFFpFp*) we can prove its mirror image as a theorem (this result is due to Lemmon); similarly with the 'density' axiom *CGGpGp* (= *CFpFFp*). In the latter case, the proof is as follows:

1. *CGGPpGPp* (*CGGpGp*, *p*/*Pp*)
2. *CGpGPp* (*CpGPp*, RGC; 1; Syll)
3. *CFHpFp* (2 *p*/*Np*, *CCpqCNpNq*, *p* = *NNp*, *NGN* = *F*, *NPN* = *H*)
4. *CFHHpFHp* (3, *p*/*Hp*)
5. *CFHHpp* (4, *CFHpp*)
6. *CHFHHpHp* (6, RHC)
7. *CHHpHp* (*CpHFp*, *p*/*HHp*; 6; Syll).

On the other hand the axioms for non-ending and for non-branching are independent of their mirror images, and vice versa.

On the whole these last results are ones we should expect; if a relation is transitive, for example, it follows that its converse is transitive, and if a series ordered by a certain relation is dense, so is the series ordered by the converse relation; but if the series ordered by a relation has a first term, it does not follow that the one ordered by the converse relation has. Linearity, however, presents a slight problem. In 'The Syntax of Time-Distinctions' the linearity of time was taken to be expressed by the 'law of trichotomy' for the earlier-later relation, *AAIabUabUba*, and from this we obviously get the same law for the converse relation. This law, however, is a stronger one than we need to express non-branching from earlier to later; for that we need only the conditional principle

CUabCUacAAIbcUbcUcb,

and this does not entail non-branching going backwards, which would be

CUbaCUcaAAIbcUbcUcb.

One other discovery of the Californian tense-logicians is that it makes a difference to one's tense-logic whether time is conceived as merely dense (like the series of rational numbers) or strictly continuous (like the reals). This was first noticed by Richard Montague, working with Cocchiarella. The difference between the logic of merely dense and that of strictly continuous time will be discussed in the course of the next chapter.

8. *Correction of Hume on past and future.* Before going on to that, it is worth making one philosophical point. J. F. Bennett recently described Leibniz as having discovered, and Hume as having re-discovered, the principle that 'if *Q* is an immediate consequence of *P* then there cannot be a time-reference in *Q* later than the latest time-reference in *P*'.[1] One thing that the development of tense-logic makes quite clear—if it was not clear before—is that this alleged 'discovery' is in fact a falsehood (consider, e.g., the law *CpGPp*, 'What is so will-always have been so'). And it *was* clear before—as usual, to McTaggart. The point arises where McTaggart is discussing an early theory of C. D. Broad's that the passage of time, or 'absolute becoming', consists in the adding of more and more layers on to the totality of 'fact'. The past and present belong to this totality, but not the future; and from this Broad deduces that there are no facts for propositions or judgements about the future to accord or discord with, so that such propositions or judgements are, strictly speaking, neither true nor false. 'Dr. Broad's theory must be false', McTaggart comments, 'if the past ever intrinsically determines the future', i.e. entails truths about it. 'If *X* intrinsically determines a subsequent *Y*, then (at any rate as soon as *X* is present or past, and therefore, on Dr. Broad's theory, real) there must be a subsequent *Y*. . . . And if that *Y* is not itself present or past, then it is true that there will be a future *Y*, and so something is true about the future.' That the past *does* sometimes 'intrinsically determine' the future, McTaggart shows by some examples, of which the simplest is that 'if Smith has already died childless, this intrinsically determines that no future event will be a marriage of one of Smith's grandchildren.'[2] Bennett's reference to Hume is of course to those passages[3] in

[1] Jonathan Bennett, 'A Myth about Necessity', *Analysis*, vol. 21, no. 3 (Jan. 1961), pp. 59–63.

[2] *The Nature of Existence*, ch. xxxiii, § 337–8.

[3] They are mostly to be found in the *Enquiry concerning Human Understanding*, Section IV.

which he denies that we have any rational basis for our supposition 'that the future will resemble the past'. 'Let the course of things be allowed hitherto ever so regular; that alone . . . proves not that, for the future, it will continue so.' He perhaps comes closest to the principle enunciated by Bennett when he says that 'past *Experience* can be allowed to give *direct* and *certain* information of those precise objects only, and that precise period of time, which fell under its cognizance', and there is no rational justification for extending this experience to 'future times' and 'other objects'. And one suspects that he *would* have assented to Bennett's principle if that had been put to him; but all that his argument really requires is something much less sweeping, namely that *CPpFp*, and even *CHPpFp*, are not laws of any normal tense-logic.

IV

NON-STANDARD TENSE-LOGICS

1. *Theses which assume that time is either discrete or circular.* LOGICAL purity, at least if one has departed from it so far as to have a tense-logic at all, is something of a will-o'-the-wisp. The logician must be rather like a lawyer—not in Toulmin's sense,[1] that of reasoning less rigorously than a mathematician—but in the sense that he is there to give the metaphysician, perhaps even the physicist, the tense-logic that he wants, provided that it be consistent. He must tell his client what the consequences of a given choice will be (e.g. that without denseness, infinity, and linearity you don't get the Hamblin reductions), and what alternatives are open to him; but I doubt whether he can, *qua* logician, do more. We must develop, in fact, alternative tense-logics, rather like alternative geometries; though this is not to deny that the question of what sort of time we actually live in, like the question of what sort of space we actually live in, is a real one, or that the logician's exploration of the alternatives can help one to decide it. It is, anyhow, worth seeing what the logic of discrete time, finite time, branching time, circular time, etc., are like, and also how far we can go without committing ourselves on this issue or that. This is the direction which the investigations of Hamblin, Scott, and Lemmon, for example, have now taken.

We may begin by glancing at discrete time. One thesis that has been associated with discrete time is the Diodorean *CKpGpPGp*, or *CpCGpPGp*, discussed in the preceding chapter. Another is *CKFNpFGpFKNpGp*, 'If both it will be that not *p*, and it will be that (it will always be that *p*), then it will be that both (not-*p* now, and *p* for ever after)', i.e. if *p* hasn't yet stopped being false but sooner or later is going to, there will be a last moment of its falsehood. This was used verbally in our informal proof of Dummett's formula *CLCLCpLppCMLpp* in Chapter 2;

[1] S. E. Toulmin, *The Uses of Argument.*

if the symbolic form is appended to, say, Cocchiarella's tense logic (from which *CFpFFp* is absent), and *M* and *L* are defined in the Diodorean way, the Dummett formula is provable. As a first step, we prove

CFGpCLCNpMKpFNpp.

We make use, where we wish to, of the fact that the laws for the Diodorean *L* and *M* defined in Cocchiarella's system are those of S4.3, and it will be useful to introduce the form *Yp* ('*p* is true for the last time') as short for *KpNFp*. This gives

YNp = *KNpGp*
NYNp = *NKNpGp* = *CNpNGp* = *CNpFNp* (= *CGpp*).

The proof (quite closely reproducing our informal one) is as follows:

C (1) *FGp*
C (2) *LCNpMKpFNp*
K (3) *LCNpMFNp* (2, *LCMKpqMq*)
K (4) *LCNpAFNpFFNp* (3, Df. *M*)
K (5) *LCNpFNp* (4, *LCFFpFp*)
K (6) *CNpFNp* (5, *CLpp*)
K (7) *CNpKFNpFGp* (1, 6)
K (8) *CNpFKNpGp* (= *CNpFYNp*), from 7 and the discreteness thesis *CKFNpFGpFKNpGp*
K (9) *GCNpFNp* = *GNYNp* (5, *CLpGp*)
K (10) *NFYNp* (9, *GN* = *NF*)
(11) *p* (8, 10, *CCNpqCNqp*).

From this we proceed as follows:

1. *CFGpCLCNpMKpFNpp* (just proved)
2. *CFKpGpCLCNpMKpFNpp* (1, *CFKpqFq*)
3. *CKpGpCLCNpMKpFNpp* (*CKpqCrp*, subst.)
4. *CAKpGpFKpGpCLCNpMKpFNpp* (3, 2, *CCprCCqrCApqr*)
5. *CMLpCLCNpMKpFNpp* (4, Df. *M*, Df. *L*).

This, as was noted in Chapter II, is equivalent to the Dummett formula. Cocchiarella's system plus *CKFNpFGpFKNpGp* thus yields, as its Diodorean-modal fragment, the system which Bull has shown to be complete for discrete Diodorean modality.

This last is more than can be said for the formula *CpCGpPGp*. For this (with Dff. *M*, *L*) yields no more than S4.3 even when rather freakishly combined with *CPpPPp*, as in Hamblin's original system. In this system we can carry out the following derivation:

1. *CpCGpPGp*	
2. *CpCGpPPGp*	(1, *CPpPPp*)
3. *CpCGpPp*	(2, *CPGpp*, RPC)
4. *CGpCGGpPGp*	(3, subst.)
5. *CGpPGp*	(4, *CGpGGp*, *CCpCqrCCpqCpr*)
6. *CGpp*	(5, *CPGpp*)

Since ⊢*CpFp* is easily deducible from ⊢*CGpp* (⊢*CGpp* → ⊢*CGNpNp* = ⊢*CpNGNp*), and ⊢*CGpFp* follows from both, this deduction shows that the axiom H1 is superfluous in Hamblin's original system. Moreover, *CGpp*, or the equivalent *CpFp*, may *replace* *CpCGpPGp*, or the equivalent *CGPpApPp*, in that system, since we can then carry out the following deduction:

1. *CpFp*	
2. *CpPp*	(1, MI)
3. *CGpPGp*	(2, subst.)
4. *CpCGpPGp*	(3, *CqCpq*).

This suggests the following more compact axiomatization of a system equivalent to Hamblin's: Take *G* and *H* as primitive, define *F* as *NGN* and *P* as *NHN*, and add to propositional calculus (with substitution and detachment) the rule RG to infer ⊢*G*α from ⊢α, the mirror-image rule, and the following axioms:

A1. *CGCpqCGpGq*	
A2. *CGpp*	
A3. *CGpGGp*	(the converse follows from A2)
A4. *CpGPp*	
A5. *CGpCHpGHp*	(initial *Cp* superfluous by A2).

If we drop the mirror-image rule and lay down mirror images separately, we need not bother to do this with A2 and A3. That A3 (in the presence of RG, A1, A4, and their images) entails its own image, has been already mentioned; that A2 does so, we show thus:

1. *CGPpPp* (A2, *p*/*Pp*)
2. *CpPp* (A4, 1, syll)
3. *CHpp* (2, *p*/*Np*; *CCNpqCNqp*; Df. *P*).

The formulae *CGpp* ('What will always be so, is so now') and *CpFp* ('What is so will be so'), and their images, are ones which would hold if time were circular. They express the reflexiveness of the earlier-later relation in circular time (i.e. everything being earlier than itself); and the reflexiveness of any relation entails that of its converse (the *U*-counterpart of the proof of *CHpp* from *CGpp*). But not all theses which would hold in circular time are provable in this system; e.g. *CGpHp* and *CFpPp* are not so provable. For all the postulates of this system are satisfied if we read *G* as 'It is and always will be' and *H* as 'It is and always has been', i.e. Diodorean 'necessity' and its mirror image, but it is easy to find counter-examples to *CGpHp* in *this* sense, i.e. to 'Whatever is and always will be so, is and always has been so.' The system is, I suspect, complete for this interpretation; at least it contains all the laws of S4.3 with *G* for *L* (since Scott's result means that we can get *CKFpFqAAFKpq-FKpFqFKqFp*, and this with *CpFp* gives *CKFpFqAFKpFqFKqFp*, i.e. Hintikka's law); and similarly of course for *H*. It is interesting that with this *G* and *H* there are the same 15 affirmative 'tenses' or 'modalities' as there are with *G* and *H* more normally interpreted—no more and no less, and with the same implication lines, except that the main diagonals go *Hp* → *p* → *Fp* and *Gp* → *p* → *Pp* instead of *FHp* → *p* → *HFp* and *PGp* → *p* → *Gp*, and this change makes the dotted lines superfluous. On the new interpretation, however, the result does not depend on time's being assumed to be dense, since here *CGGpGp* does not carry that implication; though neither, on this interpretation, does *CpCGpPGp* carry the opposite implication. With *G* and *H* interpreted in the new way, the difference between discrete and dense time is not expressible.

In this calculus, *Gp* and *KpGp* are equivalent (⊢*CKpGpGp* anyway, and ⊢*CGpp* → ⊢*CGpKpGp* by *CCpqCpKqp*). Hence the logic of a Diodorean *L* defined in this system will be precisely that of *G*, i.e. S4.3, and from this the Dummett formula for discrete Diodorean modality is known not to be deducible.

If we return to Hamblin's original postulates, take his

equivalence H4. *EApPpGPp*, break it up into two implications, and retain *CApPpGPp*, but replace the converse by the other formula which is normally associated with time's discreteness, *CKFNpFGpFKNpGp*, we obtain a much stronger system. Not only can we prove Dummett's formula, but we can also prove (remember that the system also contains *EFpFFp*) all that is involved in the idea of time's being circular. For we have the following deduction:

1.	*CKFNpFGpFKNpGp*	
2.	*CNFKNpGpNKFNpFGp*	(1, transp.)
3.	*CGNKGpNpNKFGpNGp*	(2, $Kpq = Kqp$, $NF = GN$, $FN = NG$)
4.	*CGCGppCFGpGp*	(3, $NKpNq = Cpq$)
5.	*CGGpGp*	
6.	*CFGGpGGp*	(5, RG; 4)
7.	*CFGpGp*	(6, $GG = G$)
8.	*CHFGpHGp*	(7, RHC)
9.	*CHGpHPGp*	(*CHpPp*, RHC, $HH = H$)
10.	*CHGpHp*	(*CPGpp*, RHC; 9)
11.	*CGpHGp*	(8, *CpHFp*)
12.	*CGpHp*	(11, 10).

And for the other consequence of time's circularity we have

13.	*CGpPGp*	(11, *CHpPp*)
14.	*CGpp*	(13, *CPGpp*).

12 makes *G* and *H* logically equivalent, and the laws of both are now those of the Lewis modal system S5; the only distinct tenses are *Gp* (= *Hp*), *p*, and *Fp* (= *Pp*).

2. *Postulates for circular time.* The above system has the mirror-image rule among its postulates, but E. J. Lemmon has observed that we do not really need this rule in order to turn *CGpHp* into an equivalence. His own very weak system K_t suffices for this, and indeed if that system be given we can deduce from any one of the following formulae all of the remainder: *CGpHp*, *CHpGp*, *CFpPp*, *CPpFp*, *CFGpp*, *CPHpp*, *CpGFp*, *CpHPp*. We have, for instance,

1.	*CGpHp*	
2.	*CFGpFHp*	(1, RFC)

3. *CFGpp* (2, *CFHpp*)
4. *CFGPpPp* (3, *p/Pp*)
5. *CFpPp* (*CpGPp*, RFC; 4)
6. *CHpGp* (5, *p/Np*; *CCpqCNqNp*; *NPN* = *H*, *NFN* = *G*)
7. *CPHpPGp* (6, RPC)
8. *CPHpp* (7, *CPGpp*)
9. *CPHFpFp* (8, *p/Fp*)
10. *CPpFp* (*CpHFp*, RPC; 9)
11. *CGpHp* (10, *p/Np*; *CCpqCNqNp*; *NFN* = *G*, *NPN* = *H*).

We could obviously have started this proof-circle equally well at 3, 5, 6, 8, or 10. K_t with one of these axioms alone, however, does not give us S5 for *G* (= *H*), but needs to be supplemented by *CGpGGp* and *CGpFp* (or *CGpp*). In the associated *U*-calculus, the symmetry of *U* (in circular time, if *a* is earlier than *b*, then *b* is earlier than *a*) suffices to give us 1, 3, 5, etc., but for *CGpGGp* we need transitivity, and for *CGpp* reflexiveness (in circular time, everything is earlier than itself).

But the simplest way to axiomatize circular time is to define *G* as *H*, or both as *L*, and use known postulates for S5 (e.g. RL, *CLCpqCLpLq*, *CLpp*, *CNLpLNLp*). There would no doubt be a certain artificiality in this, since even in circular time we can distinguish the past direction from the future one, but it is only the sort of artificiality which is equally present, for example, in systems of propositional calculus in which 'Not *p*' is defined in terms of a primitive 'neither' as 'Neither *p* nor *p*'. The circular *Gp* and *Hp* are equivalent functions in the sense that any proposition satisfying either satisfies the other, and also in the sense of being completely interchangeable *in the present calculus*, e.g. we have *NGp* and *KpNGp* where and only where we have *NHp* and *KpNHp*. But if we enriched the calculus with functions like '*x* knows that *p*', we would not want to equate '*x* knows that it hasn't happened' with '*x* knows that it won't happen' (since time might well *be* circular without everybody knowing it); just as we would not want to equate '*x* knows that not *p*' with '*x* knows that neither *p* nor *p*' (*x* might well not have worked out that equivalence). Moreover, if we enrich our calculus merely with new types of tense-operators, namely ones containing a reference to specific intervals, we can distinguish even in circular time between happening such and such a time hence and happening *that* time ago (we shall return to this point in Chapter VI).

We also distinguish *P* and *F* even in circular time if we adopt the rather different convention about them that was mentioned in the last chapter as giving a non-transitive earlier–later relation; i.e. the convention according to which we *don't* call a thing future or past if it's so far round the circle as to be closer to us the other way. Hamblin calls this an 'east–west' tense-logic, 'in the sense in which California is east but not west of Sydney, and west but not east of Manchester'. He has pointed out that we have some further choices here about how we shall treat 'antipodal' moments, if there be such. If there were, for example, just three moments, arranged thus:

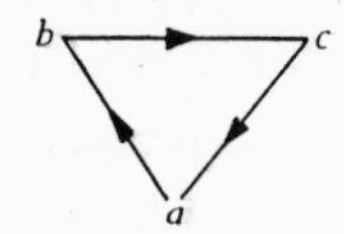

we might take *a*'s future to extend as far as *c*, or only as far as *b*; if there were four, arranged thus:

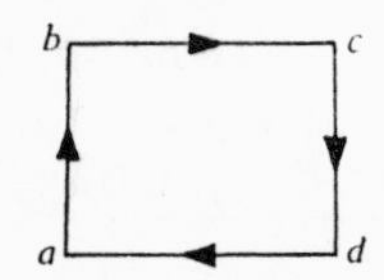

we might regard *c* as being both in *a*'s past and in its future or as being in neither (though on either view it is in the future of *a*'s future). With a dense infinity of moments there are similar but more complicated choices. In general, if we do *not* allow antipodal moments to be both past and future, we will have as a law *CFGpPp*; while if we do, we will have *CGpPp*. Given Lemmon's minimal system K_t, *CFGpPp* is deductively equivalent to each of the theses *CGGpHp*, *CFGGpp*, and to each of their mirror images and duals, and its own. We have, e.g.

1. *CFGpPp*
2. *CFGGpPGp* (1, *p*/*Gp*)
3. *CFGGpp* (2, *CPGpp*)
4. *CHFGGpHp* (3, RHC)

5. *CGGpHp* (4, *CpHFp*)
6. *CPpFFp* (5, p/Np, *CCpqCNqNp*, $NNp = p$, $NHN = P$, $NGN = F$)
7. *CPHpFFHp* (6, p/Hp)
8. *CPHpFp* (7, *CFHpp*).

8 is the mirror image of 1, and from it we can work back to 1 through the mirror images of 2–7. The axiom *CGpPp* for the other system is deductively equivalent to *CGGpp*, and to the mirror image and dual of that, and its own mirror image and dual (cf. the inter-deducible axioms for circularity taken the other way). With respect to K_t, *CFGpPp* and *CGpPp* are mutually independent (as the triangular example shows), and we may deduce *CGpFp* from the latter but not from the former—not from the former, because K_t+*CFGpPp* is consistent with reading both *Gp* and *Hp* as *Cpp*, an interpretation which rejects *CGpNGNp* (i.e. *CCppNCNpNp*); but from the latter, as follows:

1. *CKGpGqGKpq* (provable in K_t)
2. *CKGpGNpGKpNp* (1 q/Np)
3. *CGKpNpPKpNp* (*CGpPp*)
4. *CPKpNpNHNKpNp* (Df. *P*)
5. *CKGpGNpNHNKpNp* (2, 3, 4, syll.)
6. *HNKpNp* (*NKpNp*, RH)
7. *NKGpGNp* (5, 6, *CCpNqCqNp*)
8. *CGpNGNp* (7, *CNKpqCpNq*).

3. *Postulates for the next and the last moment, in discrete time.* For a logic of discrete but *not* necessarily circular time we need to add some appropriate axiom to a form of tense-logic, e.g. Scott's or Cocchiarella's, which does not assert *CGGpGp*. Perhaps *CKFNpFGpFKNpGp* would be sufficient, though this is not known. Scott has proved completeness for a slightly different type of system for discrete time, one in which *G* and *H* are not the only undefined tense-symbols, but are supplemented by two others for 'It will be true at the next instant that *p*' and 'It was true at the last instant that *p*'. We could write *Tp* and *Yp* for these new forms (after 'tomorrow' and 'yesterday'). They are not definable in terms of *G* and *H*. We can, indeed, define '*p* will be true at the next instant *for the last time*' as *KFpNFFp*, '*p* will be true at the next instant but one for the last time' as

KFFpNFFFp, etc.; but the simple '*p* will be true in the next instant', '—in the next but one', etc. (i.e. *Tp*, *TTp*, etc.) cannot be quite got this way. Conversely, we cannot hatch *G*, *H*, *F*, or *P*, out of *T* and *Y*; only more specific forms of *Fp* like *Tp*, *ATpTTp*, *AATpTTpTTTp*, etc., and finite approaches to *Gp* like *Tp*, *KTpTTp*, etc. So Scott takes both pairs (*G*, *H*, and *T*, *Y*) as primitive, and has proved completeness for the system in which his *GH* postulates are supplemented by the following axioms:

T1. *CGpTp* Y1. *CHpYp*
T2. *ENTNpTp* Y2. *ENYNpYp*
T3. *CTCpqCTpTq* Y3. *CYCpqCYpYq*
T4. *CpYTp* Y4. *CpTYp*
T5. *CTpCGCpTpGp* Y5. *CYpCHCpYpHp*.

The last pair are 'inductive' axioms; the first states that if *p* will be true at the next instant then if it will always be that if *p* is true it is true the instant after, then *p* will be true for ever; ditto for 'has been' in the other.

The usefulness of systems of this sort does not depend on any serious metaphysical assumption that time *is* discrete; they are applicable in limited fields of discourse in which we are concerned only with what happens next in a sequence of discrete states, e.g. in the workings of a digital computer.[1]

Scott's system for *G*, *H*, *Y*, *T* appears to have been developed from one which he had devised by 1964 (proving its completeness) in which the following postulates for *T* and *Y* were subjoined to Gödel's axiomatization of S5 for *L* ('at all times'):

1. *ELpTLp* 2. *ELpLTp*
3. *ETNpNTp* 4. *ETCpqCTpTq*
5. *ETYpp*
6. *CLCpTpCLCqYqCMKpqLApq*

The last 'inductive' axiom says that if *p*-now always implies *p*-next, and *q*-now always implies *q*-just-past, and at some time both *p* and *q*, then at all times either *p* or *q* (since from the time at which both, we will have *p* all the way forward, step by step, and *q* all the way back, step by step). Of the other

[1] Cf. H. Greniewski, K. Bochenek, R. Marczyński, 'Application of Bi-elemental Boolean Algebra to Electronic Circuits', *Studia Logica*, ii (1955), pp. 7–74.

axioms, at least 4 may be replaced by the corresponding implication. For, firstly, the rule to infer $\vdash T\alpha$ from α may be established as follows:

$$\begin{aligned} \vdash\alpha &\rightarrow \vdash L\alpha && \text{(RL, from S5)} \\ &\rightarrow \vdash LT\alpha && \text{(by 2)} \\ &\rightarrow \vdash T\alpha && \text{(by } CLpp\text{, from S5).} \end{aligned}$$

This, with *CTCpqCTpTq*, yields the rule

$$\text{RTC: } \vdash C\alpha\beta \rightarrow \vdash CT\alpha T\beta.$$

We then have

7.	*CTKpNqTp*	(p.c., RTC)
8.	*CTKpNqTNq*	(p.c., RTC)
9.	*CTKpNqKTpTNq*	(7, 8, p.c.)
10.	*CTNCpqTKpNq*	(p.c., RTC)
11.	*CKTpTNqKTpNTq*	(p.c., 3)
12.	*CTNCpqKTpNTq*	(10, 9, 11, Syll)
13.	*CTNCpqNCTpTq*	(12, p.c.)
14.	*CNTCpqNCTpTq*	(13, 3)
15.	*CCTpTqTCpq*	(14, p.c.).

This is the converse implication that makes up the rest of 4. (This adapts a proof given by Rescher in another connexion.) From 4 in turn we can get

16. *ETEpqETpTq*,

and from this the rule that if $\vdash ET\alpha T\beta$ then $\vdash E\alpha\beta$, thus:

$$\begin{aligned} \vdash ET\alpha T\beta &\rightarrow \vdash TE\alpha\beta && \text{(by 16)} \\ &\rightarrow \vdash LTE\alpha\beta && \text{(by RL)} \\ &\rightarrow \vdash LE\alpha\beta && \text{(by 2)} \\ &\rightarrow \vdash E\alpha\beta && \text{(by } CLpp\text{).} \end{aligned}$$

This rule—we may call it RET—is useful in proving the mirror images of the axioms; e.g.

17.	*ETYTpTp*	(5, *p*/*Tp*)
18.	*EYTpp*	(17, RET).

This system suggested to Lemmon (in 1964) the following for the future only, with *G* read as 'It is and always will be' (Diodorean necessity):

RG:— $\vdash\alpha \rightarrow \vdash G\alpha$

A1. *CGpp* A2. *CGCpqCGpGq*
A3. *ETNpNTp* A4. *ETCpqCTpTq*
A5. *ETGpGTp* A6. *CGpTGp*
A7. *CGCpTpCpGp*.

From these he proved S4.3 plus the Dummett discreteness formula in *G*; and from these for *G* and *T*, plus an analogous set for *H* and *Υ*, the mixing axioms *ETΥpp* and *EΥTpp*, and the definition of Scott's *L* as *KGpHp*, he proved Scott's *L–Υ–T* postulates. Since the ordinary *Gp* is equivalent in this system to *TGp*, Scott's 1965 system ought also to be capable of development within this one.

Scott's 1965 system does not separate off the theses in *G* and *H* only which assume discrete time; nor do any of these systems separate off from the rest the pure logic of *T* and *Υ*. The axiomatization of the logic of *T* alone, however, has been solved by G. H. von Wright and J. Clifford.[1] The postulates for which von Wright establishes completeness are not directly formulated in terms of Scott's *T* but in terms of another *T*, a dyadic operator such that the form *Tpq* may be read as '*p* now and *q* next' (i.e. in the next state or instant). But as Clifford points out, von Wright's *T* is definable in terms of Scott's, by $T_W pq = KpT_S q$; and Scott's can be defined in terms of von Wright's, by $T_S p = T_W Cppp$ ('*p* next' = 'if-*p*-then-*p* now and *p* next'); so that they cover precisely the same area. Von Wright subjoins to the propositional calculus with substitution and detachment the rule of extensionality for *T* (from $\vdash E\alpha\beta$ to infer $\vdash Ef\alpha f\beta$, where *f* is any function of propositions in the system) and the following four axioms:

A1. *ETApqArsAAATprTpsTqrTqs* ('Distributivity')
A2. *EKTpqTrsTKprKqs* ('Co-ordination')
A3. *EpTpAqNq* ('Redundancy'; cf. the definition of Scott's *T* above.)
A4. *NTpKqNq* ('Impossibility').

The rule of extensionality could perhaps be replaced by something less comprehensive, e.g. the pair $\vdash E\alpha\beta \rightarrow \vdash ET\alpha\gamma T\beta\gamma$ and

[1] G. H. von Wright, '"And Next"', *Acta Philosophica Fennica*, Fasc. 18 (1965), pp. 293–304; J. Clifford, 'Tense logic and the logic of change' (revision of the Rudolf Carnap Prize essay at the University of California in Los Angeles, 1965), *Logique et Analyse*, No. 34 (June 1966), pp. 219–30.

$\vdash E\alpha\beta \rightarrow \vdash ET\gamma\alpha T\gamma\beta$. Clifford shows that von Wright's system is derivable, given the definition of his *T*, from the axioms 1. *CTNpNTp*, 2. *CNTpTNp*, and 3. *CTCpqCTpTq*, and the rule to infer $\vdash T\alpha$ from $\vdash\alpha$. Here 1 and 2 are equivalent to Scott's T2, and 3 is Scott's T3, so Clifford's axioms amount to the only two of Scott's in which *T* is the only tense-operator (his rule follows from Scott's RG and T1). *Υ* can be axiomatized in the same way; Clifford has also shown that the logic of the two together needs nothing beyond their separate postulates except *CpΥTp* and *CpTΥp*.

4. *The logic of 'and then'*. Von Wright's 'And Next' system is a development of the logic of change sketched in his *Norm and Action*.[1] Miss Anscombe has a related logic, not of 'and next', but of 'and then', or more precisely 'It was the case that *p* and then it was the case that *q*'.[2] This *Tpq* is not definable in terms of Scott's *T*, but it is definable in terms of the *P* of ordinary tense-logic, as *PKPpq*, 'It has been the case that (it has been the case that *p*, and now it is the case that *q*)'. The converse definability is also possible if we take time to be non-discrete, since we can then define *Pp* as *TpCpp*, 'It was the case that *p* and then it was the case that if-*p*-then-*p*' (or any other thing that is true at all times). In discrete time these are not quite equivalent, for if *Pp* is true because *p* has *just* been true, there is no past moment between then and now for *Cpp* to hold in. If, however, we modify the meaning of *Tpq* to *AKPpqPKPpq*, i.e. 'It *either is or has been* the case that (it has been that *p*, and it is the case that *q*)', *Pp* is equivalent to this sense of *TpCpp* in any sort of tense-logic.[3]

For Miss Anscombe's *T*, as she points out, we can establish such laws as *CTTpqrTpq*, *CTTpqrTqr* and *CTTpqrTpr*, and we can define 'It has been the case repeatedly that *p*' as *TTpNpp*, 'It has been that (*p* and then not *p* and then *p*)'. Hence even in a non-discrete tense-logic we can distinguish between '*p* has been true at least one distinct time' (*Pp*), '*p* has been true at least two distinct times' (*TTpNpp*), '*p* has been true at least

[1] *Norm and Action* (London, 1963), ch. ii.

[2] G. E. M. Anscombe, 'Before and After', *Philosophical Review*, vol. 73, no. 1 (Jan. 1964), pp. 3–24.

[3] This modification, for this purpose, was suggested independently by Geach and Cocchiarella.

three distinct times' (*TTTTpNppNpp*) and so on. This has the following consequence. In non-beginning, non-ending, non-discrete linear time there are precisely fifteen non-equivalent forms using only *P*, *H*, *F*, *G*, and a single propositional variable. Using *P*, *F*, and *N* (and defining *H* and *G*), or using *H*, *G*, and *N* (and defining *P* and *F*), there are exactly thirty such forms, the preceding and their negations. (Hamblin's theorem was in fact originally presented as one that there are exactly thirty 'tenses', *N* being allowed to enter into the definition of a 'tense'.) But if we ask how many non-equivalent forms we can construct using *P*, *F* (or *H*, *G*), *N*, and *K*, using the same propositional variable throughout each formula, the number immediately rises to infinity, since we have, for example, the series of *T*-*N* forms just mentioned. We may contrast tense-logic at this point with, say, the modal system S5, in which there are three non-equivalent forms using *L*, *M* and a variable, six using *L* and *N* (or *M* and *N*), and sixteen using *L*, *N*, and *K*.[1]

5. *Mere denseness and Dedekindian continuity.* At the opposite pole from discreteness is strict Dedekindian continuity, though at certain points this combines features of discreteness and density. It is characteristic of a merely dense series that two adjacent segments of it may each have no first and no last member. To take the stock example, the rational numbers may be divided completely into those that are less than the square root of 2 and those that are greater, but there is no largest rational that is less than the square root of 2, and no smallest one that is greater (you can always get a little closer both ways). In the real numbers, however, the square root of 2 is itself included, and there is no room among them for 'gaps' of the sort just described; hence with any two adjacent segments, either there is a largest real number in the lower segment but no smallest one in the upper, or vice versa. (If there were both a largest number in the lower and a smallest in the upper, the series would be neither continuous nor dense, but discrete.) This feature of continuous series means that they share certain more-or-less inductive principles with discrete ones, and confusion with discrete series can be excluded by combining these with

[1] Cf. R. Carnap, 'Modalities and Quantification', *Journal of Symbolic Logic*, June 1946, p. 48.

postulates for density (e.g. in tense-logic, *CGGpGp*). One such inductive principle, noted by Cocchiarella, is the following:

CGpCHGCGpPGpHGp.

In a dense series with gaps there could be counter-examples to this. Take a case where time is divided into an earlier part throughout which *p* is false and a later part throughout which it is true, but let there be no last instant of *p*'s falsehood or first instant of *p*'s truth, and let the present instant be within the period of its truth. The first antecedent *Gp*, 'It will always be that *p*', will now be true; and all the times at which *Gp* is true will be ones at which it was also true at least a little bit before, i.e. we will have the other antecedent *HGCGpPGp*; but the consequent *HGp*, asserting the truth of *p* throughout the whole of time, will be false. But if this sort of situation is precluded by its having to be the case either that there is a last time at which *p* is false or a first at which it is true, such counter-examples to Cocchiarella's formula cannot be constructed. If, for instance, there is a last moment of *p*'s falsehood, this will be a moment at which we have *Gp* but not *PGp* (refuting the antecedent *HGCGpPGp*, and so establishing the implication); while if there is a first moment of *p*'s truth, *this* will be a time at which we have *Gp* but not *PGp*, if there is no last moment of *p*'s falsehood.

6. *Postulates for beginning and ending time*. These questions concern time's *minima*; we turn now to its *maxima*. In the matter of time's infinity both ways, we may again simply drop those axioms which commit us to this (as Cocchiarella does), or lay down ones which positively preclude it. Ending and beginning time were possibilities which entered in a small way into the discussions of Diodorean modal logic in the late 1950s. Lemmon and Dummett in 1959 noticed some of the effects of taking the *Time and Modality* matrix for *D* and reversing it, i.e. using indefinitely long strings of truth-values coming back from a fixed point instead of going forward to one, but determining the values of *L* and *M* as usual. This amounted to using the Diodorean definitions in an ending time. Their main result was that this verified the formula *ELMpMLp*. This will be seen on reflection to be natural enough. *MLp* is true only if *p*

eventually reaches a point at which it is true and after which it is never false; *LMp* is distinguished from this by being also true if *p* never reaches a time after which it is *never* false, but also never reaches a time of being false which is not followed by a later time of being true. If time comes to an end, however, these two conditions coincide, both being met if and only if *p* is true at the last moment of time, no matter what goes on before that; hence, with such a picture, we have *CLMpMLp* as well as its converse. Modal systems containing *CLMpMLp*, with or without its converse, have been studied more recently by Sobociński and others.[1]

This suggests that *CGFpFGp* might be a law in ending time, but it is not; nor is anything whatever of the form *C*$G\alpha F\beta$, since at the end of time anything of the form $G\alpha$ will be true and anything of the form $F\beta$ false. The latter (that anything of the form 'It will be that β' is false at the end of time) is obvious immediately; the former (that anything of the form 'It will always be that α' is true at the end of time) depends on our understanding *G* as equivalent to *NFN*; for any α, at the end of time it will be false that α is ever going to be false. One formula that *is* a law if time has an end is *AGpFGp*, or to give a more intuitive equivalent, *ANFpFNFp*. Whatever *p* might be, the first disjunct of this (*NFp*) is bound to be true at the last moment of time (whether or not it is true before that), and therefore the other component (*FNFp*) is bound to be true *up to* the end of time (though *at* the end, it will be false); hence one or the other of them, and therefore the whole disjunction, will be true always. Of other laws which hold in more normal systems, *CGpFp* and *CGpFGp* will be true up to the last moment of time, but false at that moment, and the same is true of the non-standard principle *CGFpFGp*; the converse of the last, *CFGpGFp*, and for that matter the plain *GFp*, is true at the last moment of time only. The same things may be said, *mutatis mutandis*, if time had a beginning.

If we attempt to combine the ending-time principle *ANFpFNFp*, or *AGpFGp*, with the non-ending-time principle

[1] B. Sobociński, 'Remarks about Axiomatizations of Certain Modal Systems', *Notre Dame Journal of Formal Logic*, vol. 5, no. 1 (Jan. 1964), pp. 71–80; A. N. Prior, 'K1, K2 and Related Modal Systems', ibid., vol. 5, no. 4 (Oct. 1964), pp. 299–304; B. Sobociński, 'Modal System S4.4' and 'Family K of the non-Lewis Modal Systems', ibid., pp. 305–12 and 313–18.

CGpFp, we obtain results which are not merely odd, like time being circular, but downright contradictory. For we will have:

1.	*AGpFGp*	
2.	*CGpFp*	
3.	*AFpFGp*	(1, 2, p.c.)
4.	*CNFpFGp*	(3, $A = CN$)
5.	*CNFNCppFGNCpp*	(4, $p/NCpp$)
6.	*GCpp*	(*Cpp*, RG)
7.	*FGNCpp*	(5, 6, $G = NFN$)
*8.	*NGFCpp*	(7, $FGN = NGNNFNN = NGF$)
9.	*FCpp*	(2, 6)
*10.	*GFCpp*	(9, RG).

(The auxiliary theses used here are all in the minimal tense-logic K_t.)

The law *AGpFGp* also differs from all the principles we have so far considered, whether in discrete, dense, continuous, linear, or branching time, in that it is *not* consistent with equating both *Gp* and *Fp* (and *Hp* and *Pp*) with the plain *p*. For whereas this turns, say, *CGpFp* into *Cpp*, and *CGCpqCGpGq* into *CCpqCpq*, and *CpGPp* into *Cpp*, it turns *AGpFGp* into *App*, which by *CAppp* gives the simple *p* as a thesis (and therefore by substitution anything at all as a thesis).

If we combine ending time with discreteness, using both *AGpFGp* and *CKFNpFGpFKNpGp*, we can very easily prove *CFpFKpNFp*, i.e. 'If it sooner or later will be that *p*, then it sooner or later will be that *p*-for-the-last-time'. This is an intuitively obvious consequence of this combination. Conversely, from *CFpFKpNFp* we can prove each of *AGpFGp* and *CKFNpFGpFKNpGp*.

7. *Tense-logic as giving the cash value of assertions about time.* Postulates of the sort we have been considering can be regarded as giving the *meaning* of such statements as 'time is continuous', 'time is infinite both ways', and so forth. This is different from saying that such postulates give the meaning of the expressions that occur in them, in particular of *F*, *H*, *G*, and *P*. Talk of this sort seems to me confused. Apart from any other objections,[1] if

[1] See A. N. Prior, 'The Runabout Inference-Ticket', *Analysis*, vol. 21, no. 2 (Dec. 1960), pp. 38–39; and 'Conjunction and Contonktion revisited', ibid., vol. 24, no. 6 (June 1964), pp. 191–5.

different postulates for *F*, *G*, *H*, and *P* define different meanings, then people who say that time has no end, for example, and who therefore agree to *CGpFp*, and people who say that it has an end, and who therefore agree to *AGpFGp*, are not really contradicting one another, since they are using words with different meanings and simply talking past one another. There are no such objections, however, to saying that what is meant by time's having an end is precisely that for any *p*, either already it will never be the case that *p*, or it will be the case that it will never be the case that *p* (or to put it another way, that it either is the case, or will be the case, that nothing—not even that such-and-such *has* occurred—*will be the case* any more). Or to saying that what is meant by time's being circular is precisely that for any *p* (however detailed or comprehensive), if it is or has been the case that *p*, then it will be the case that *p* again. Or to saying that what is meant by time's being dense ('continuous' in the looser sense) is precisely that if it will be the case, however soon, that *p*, then it will be the case, even sooner, *that* it will be the case that *p*. And there is some positive advantage in saying that this is the sort of thing we mean when we make remarks of this kind. For if taken literally, statements like 'Time will have an end', 'Time is circular', 'Time is continuous', etc., suggest that there is some monstrous object called Time, the parts of which are arranged in such-and-such ways (a common idea is that of a string on which events are strung like beads); and such statements cease to carry such suggestions when they are interpreted as short-hand for statements which do not even appear to mention any such entity, but simply talk about what will have been the case, etc.[1]

It is true that in our technical work, when we are deciding which formulae express discreteness, finitude, etc., we always turn to 'U-calculi' in which the terminology is decidedly more abstract, and time appears as something like a class of classes of propositions ordered by a certain relation. This in itself, however, doesn't make U-calculi more than handy diagrams; they need not be taken with any great metaphysical seriousness. Much more awkward is the fact that many of the conditions which might be put upon the relation *U* in a U-calculus are not expressible as theses in *G* and *H*. For example, although

[1] This, I take it, is the point of Wittgenstein's remark in the *Blue Book*.

symmetry, giving circular time, can be expressed by making *CGpHp* a tense-logical thesis, it does not appear that we can so express the position that time is *not* circular. But there is more to tense-logic than has so far been given, and certain enrichments of the symbolism can be expected to fill these gaps. Much can be done, for instance, simply by making explicit the quantifiers over sentential variables that are implicit in saying that something is a thesis, i.e. that *for all p*, such and such holds.[1] If we bring that into the symbolism, we can also say that for some *p*, such and such does *not* hold, e.g. that for some *p*, it will always be that *p* but has not always been that *p*; which does state non-circularity. We shall later see, indeed, that the U-calculus can be defined within a not much enlarged *GH* one.

[1] That such quantifications do not commit us to new entities I have argued elsewhere, e.g. in 'Oratio Obliqua', *Proc. Arist. Soc.*, Supplementary vol. 38 (1963), pp. 115–26.

V

THE LOGIC OF SUCCESSIVE WORLD-STATES

1. *The de-trivializing of modality: 'the world'*. SMILEY's proof of the consistency of most tense-logics (that they survive the interpretation $Gp = Hp = FP = Pp = p$) applies to most modal logics also. For example, the rule to infer $\vdash L\alpha$ from $\vdash \alpha$ becomes one to infer $\vdash \alpha$ from $\vdash \alpha$, *CLCpqCLpLq* become *CCpqCpq*, and *CLpp*, *CLpLLp* and even the S5 *CMLpLp* become *Cpp*, when *L* and *M* are thus trivialized. It is sometimes felt that while this does prove consistency, it also shows that the modal operators are insufficiently characterized by these calculi. This defect may be remedied in various ways. One may, for example, devise modal calculi for which such an interpretation is *not* possible; for example, the Lewis calculi which are sometimes called S6, S7, and S8, in which *MMp* is a thesis, cannot be so interpreted. Or one may follow Łukasiewicz and Thomas in introducing not only the 'turnstile' $\vdash$ to indicate that what follows *is* a thesis, but also the reversed turnstile $\dashv$ to indicate that what follows is *not* a thesis, and have such 'rejections' as $\dashv$*CpLp* and $\dashv$*CMpp*. Or we may—as Lewis himself does—introduce propositional quantifiers, say Πp for 'For all *p*' and Σp for 'For some *p*', and have such theses as $\Sigma pKMpNp$, 'Something is possible but not true'.[1] Or, finally, one may introduce some contingent propositional *constant*, i.e. some specific proposition *a* such that $\vdash$*KaMNa*.

The trouble with the last alternative is that it is difficult to find a contingent proposition which is of sufficient logical interest to merit a place in a logical calculus. C. A. Meredith[2]

[1] An extension of S5 of this sort is touched upon at the end of Saul A. Kripke's 'A Completeness Theorem in Modal Logic', *Journal of Symbolic Logic*, vol. 24, no. 1 (March 1959), pp. 1–14.

[2] C. A. Meredith and A. N. Prior, 'Modal Logic with Functorial Variables and a Contingent Constant', *Notre Dame Journal of Formal Logic*, vol. 6, no. 2 (April 1965), pp. 99–109.

has suggested that one logically interesting contingent proposition is 'the world' as defined in the first sentence of Wittgenstein's *Tractatus*—'everything that is the case'. For this 'sum of all truth', Meredith introduces the symbol *n*, with the axioms

1. *n* ('the world is the case')
2. *CpLCnp* ('the world is *everything* that is the case')
3. *CLnp* ('the world is not necessary').

Literally, 2 says that if it is the case that *p*, then 'the world' necessarily implies that *p*; and 3 that if 'the world' is necessary, anything at all is the case. The more straightforward *NLn* would do here, but Meredith's variant makes it possible to define *N* in terms of *n*. In a modal calculus with these axioms subjoined, the rule to infer $\vdash L\alpha$ from $\vdash \alpha$ will not hold, since *n* is a thesis but *Ln* is anything but; Meredith therefore subjoins it to a modal logic which does not have this rule, though it contains the same theses in *C*, *N*, and *L* as Lewis's S5. This calculus will not survive the translation of *Lp* as *p*, and for consistency Meredith gives a 4-valued matrix, which can be interpreted by supposing that there are only two possible worlds, *n* (the actual one) and $\bar{n}$, and the four 'values' that propositions can take are 'true in both worlds' (i.e. necessarily true), 'true in the actual world only' (i.e. contingently true), 'true in the alternative world only' (i.e. contingently false) and 'false in both worlds' (i.e. necessarily false), and the laws are those formulae which always hold (i.e. for all values of their variables) either in this world only or in both. The matrix is

	C	1	*n*	$\bar{n}$	0	*L*
*	1	1	*n*	$\bar{n}$	0	1
*	*n*	1	1	$\bar{n}$	$\bar{n}$	0
	$\bar{n}$	1	*n*	1	*n*	0
	0	1	1	1	1	0

It has been pointed out by R. Suszko that this solution to the problem of preventing confusion between *Lp*, *Mp*, and *p* can be assimilated to the preceding one by dropping Meredith's constant, and introducing instead a function *Wp* which asserts in effect that *p* has the properties of that constant, i.e. that *p* is a truth so comprehensive that all other truths follow from it. With propositional quantifiers, we can define *Wp* as *KpΠqCqLCpq*, '*p* is a truth, and for any *q*, if *q* is true then *p* necessarily implies it'. This immediately gives us Meredith's first two axioms under a condition, i.e. we get *CWpp* and *CWpCqLCpq*. For the third, we need to lay it down that there is at least one contingent truth, *ΣpKpMNp*, from which it follows that the totality of truth is contingent, *CWpMNp* (or *CWpNLp*, or *CWpCLpq*). This procedure has the advantage of not committing us to the view that there are in fact any such all-comprehensive propositions, still less that there is exactly one of them; though we can lay down *ΣpWp* as a further axiom if we wish to, and we can easily prove that if *p* and *q* are both all-comprehensive truths they are necessarily equivalent, *CWpCWqLEpq*.

2. *Instantaneous world-states*. There are the same different possible solutions to the problem of precluding the trivialisation of tense-logic. We may, on the one hand, adopt a non-standard tense-logic which will not survive Smiley's translation of the symbols, e.g. the one for an ending time with ⊢*AGpFGp*. Or we may introduce a rejection sign and put it before, say, *CpFp*. Or we may introduce propositional quantifiers and introduce such axioms as *ΣpKpFNp*, 'Something is now true which will be false'. Or we may introduce a constant for a proposition that expresses the total present state of the world, with axioms similar to Meredith's. Or we may introduce a function *Wp* which means that *p* is a present truth from which everything that is now true permanently follows, i.e. *KpΠqCqLCpq*, where $L\alpha = KK\alpha H\alpha G\alpha$, or if you like $L = GH$. Note carefully what the last part of this definition says; it means that if *p* expresses the total present world-state, and *q* is now true, then although both *p* and *q* may be false at other times (and also *p* may be false and *q* still true), the relation between them is such—*p* so *contains q*—that the implication of *q* by *p* will be true even at those other times, in fact at all times, however the world changes. (The now-true proposition *q* need not of course always be implied at other times by what is *then* the totality of truth, and if it is false it won't be, but it will be implied even then by what is now the totality of truth.)

It is also possible to drop the propositional quantifiers and simply introduce the function Wp as a new primitive, laying down for it the axioms

$$\text{W1. } CWpp$$
$$\text{W2. } CWpCqLCpq.$$

In making a few deductions from these postulates, it will be simplest to subjoin them, in the first instance, to a tense-logic strong enough to yield for L, defined as above (and M analogously, or as NLN), the Lewis system S5. We may begin with the theorem that if p gives the total present world-state it is permanently equivalent to the assertion *that* it gives the total present world state, $CWpLEpWp$. For whenever p is the totality of what is then true, it is true ($LCWpp$), and if p permanently implies whatever is true, it permanently implies that it does so ($LCpWp$). Moreover, not only if now, but if *at any time* p is the totality of what is then true, it is permanently equivalent to the statement that it is that. This follows from the preceding result, thus:

T1.	$CWpLEpWp$	(just proved)
T2.	$LCWpLEpWp$	(T1, RL)
T3.	$CMWpMLEpWp$	(T2, $CLCpqCMpMq$)
T4.	$CMWpLEpWp$	(T3, $ML = L$).

Another theorem is that if at any time p is the totality of what is then true, then whatever q may be, either p permanently implies q or p permanently implies not q. Proof:

W2.	$CWpCqLCpq$	
T5.	$CWpCNqLCpNq$	(W2, subst.)
T6.	$CWpCAqNqALCpqLCpNq$	(W, T5, $CCpCqsCCpCrtCpCAqrAst$)
T7.	$CWpALCpqLCpNq$	(T6, $AqNq$)
T8.	$CMWpMALCpqLCpNq$	(T7, RL, $CLCpqCMpMq$)
T9.	$CMWpAMLCpqMLCpNq$	(T8, $MApq = AMpMq$)
T10.	$CMWpALCpqLCpNq$	(T9, $ML = L$).

We might in fact have begun by defining a form Qp meaning 'p is the totality of truth at some time', i.e. is a 'possible world' in the present sense of 'possible', as: $KMp\Pi qALCpqLCpNq$; and then defined Wp as $KpQp$. The simple $\Pi qALCpqLCpNq$ is true not only of 'worlds' but also of impossibilities, i.e. (in this context), what is never true, since these permanently imply *all* propositions, so we might have defined a form *Op*, meaning '*p* is either a world or an impossibility', as *ΠqALCpqLCpNq*, and then defined *Qp* as *KMpOp*. The separate logic of *O* and *Q*, especially of *Q*, is worth investigating, but here we will simply prove occasional theses about the equivalent *MW*. The ordinary modal variant of *Op* corresponds to what Carnap calls '*L*-completeness', and that of *Qp* to what he calls being an '*L*-state', though in Carnap these are relativized to a language.[1]

In general the negation of a logically strong proposition is a comparatively weak one, e.g. to contradict an Aristotelian universal form 'Every *S* is *P*' we don't need to assert the equally 'extreme' proposition 'No *S* is *P*' but only the comparatively mild 'Some *S* is not *P*'. It might therefore appear that in order to contradict so immense an assertion as the totality of truth we need only say something very feeble, which cannot possibly itself be a 'world-proposition'; so that there ought to be a theorem that if *p* is a 'world-proposition', 'not *p*' is not, *CMWpNMWNp* (= *CQpNQNp*). This cannot, however, be proved from the basis given. What *can* be proved is that if *p* and *Np* are both world-propositions, they are the *only* world propositions, at least in the sense that every world-proposition is permanently equivalent to one or the other of them. We may symbolize this as *CMWpCMWNpLAWpWNp*, and prove it as follows:

C	(1)	*MWp*	
C	(2)	*MWNp*	
K	(3)	*LEWpp*	(1, T4)
K	(4)	*LEWNpNp*	(2, T4)
K	(5)	*LApNp*	(*ApNp*, RL)
	(6)	*LAWpWNp*	(3, 4, 5).

We can also prove a kind of converse of this, namely that if there are exactly two world-states, each is permanently equivalent to the negation of the other; i.e. if there are at most two world-states, then if they are not one and the same (or permanently equivalent), each is equivalent to the other's negation, *CLAWpWqCNLEpqLEpNq*. From this it follows that if a world-state is always either *p* or *q*, it is always either *p* or *Not-p*,

[1] R. Carnap, *Introduction to Semantics* (1941), pp. 94, 107.

CLAWpWqLAWpWNp. (Cf. the *n* and *ñ* of Meredith's consistency-matrix.)

3. *The logic of 'worlds' and Laplacean determinism.* One proposition which follows immediately from W2 (by substituting *FWq* for *p*) is

(A) *CWpCFWqLCpFWq*,

'If *p* gives the present total world-state, then if *q* is a future total world-state, *p* permanently-implies that *q* is a future total world-state'. All future world-states, in other worlds, are implied by the present one. It would be pleasant (or disastrous, according to taste) if we could use this as a logical proof of Laplacean determinism; but putting it to that use would be cheating. For *this* 'totality of present truth' is understood as including all such future-tense propositions as are true now, including such truths as there may now be about what the future world-states are; Laplacean determinism, I think, asserts the deducibility of the future from rather more restricted premisses, or perhaps claims that the 'totality of present truth' in our sense is deducible from a set of propositions giving (*a*) the totality of 'present' truth in a more restricted sense, and (*b*) certain permanent natural laws. That proposition (A) is not itself Laplacean appears plainly from the fact that it still holds within a kind of tense-logic which is quite un-Laplacean, namely one without a future-tense linearity axiom, in which there are alternative futures. In this system, it may be remembered, *KKpHpGp* (*Lp*) does not imply *HGp* (though it does imply *GHp*), and it might be thought to strengthen *W2* if we replaced *L* there by *HG*, i.e. if we read it as *CWpCqHGCpq*. But this alteration would still not make *W2* deterministic. For the only information about the future that is conveyed by *FWq* in a branching time system is that *q* gives one of the momentary total world-states in some *possible* future course of events, and all that the present total world-state permanently implies (i.e. all that follows whenever the total world state is as it is now) is that *q could* be a future world state. This is certainly less than the Laplacean theory.

4. *Non-repetition, repetition, and world-state-hood.* One point at which it does make a difference (in branching time) whether

we use *GH* or *HG* to define the *L* in the definition of *Wp* as *KpΠqCqLCpq*, is the following: If we use *GH*, we can prove that if *p* is a proposition true at the present time only, then for such a *p* we have *Wp*, in the sense of this definition, i.e. besides being true it will permanently imply ('materially imply', of course) whatever is now true. What we have to prove here is *CKKpHNpGNpΠqCqGHCpq*, and the proof is as follows:

ΠqC (1) *p*
  C (2) *HNp*
  C (3) *GNp*
  C (4) *q*
  K (5) *Cpq* (4, *CqCpq*)
  K (6) *HCpq* (*CNpCpq*, RHC; 2)
  K (7) *GCpq* (*CNpCpq*, RGC; 3)
   (8) *GHCpq* (4, 5, 6, *CKKpHpGpGHp*).

(Informally: now, when *q* is true, it is materially implied by anything, e.g. by *p*; at all other times, when *p* is false, *p* materially implies anything, such as *q*.) This oddity reflects the difference between merely *permanent* implication and *logical* implication. The theorem can also be proved with an *M* before both antecedent and consequent, i.e. if *p* is or has been or will be true at one time only, then it is *a* 'world' proposition in the sense defined, though not necessarily the present one. (We get this result directly from the last by RMC.) And in a linear time-scheme, in which we have the mirror image of the thesis used in proving line (8) above, we can also prove line (8) with *HG* for *GH*. But in a time-scheme with alternative futures, in which we do not have this mirror image, this proof will fail. Intuitively, what *KKpHNpGNp* will now mean is that *p* is now true, and as it happens has always been false (but might not have been) and is bound to be false for all time hereafter. It is as if it had never until now tried being true, and having tried it once, is scared off it for all possible future time. (We can drop this anthropomorphism, if it's worrying, by supposing it to be a proposition about an individual behaving like this with respect to some particular thing.) And it might in fact have tried it in the past *without* being scared off, and then repeated it under different circumstances, and so *not* have permanently-materially-implied all the circumstances of its original occurrence; this unfits it for being a 'world' proposition with W defined in terms of HG.

Wp does not conversely entail that p is true at one time only; but because of the comprehensiveness (however 'extensional') attributed to p by Wp, in the sense of W which would justify W2, a number of theorems can be proved about the *repetition* of total world states, if any such thing should occur. For example, it can be proved that if 'we have had all this before' (*all* of it), we'll have it all again, $CWpCPWpFWp$. Intuitively, the point is simple. If we had before the same totality of truth that we have now, then part of what we had *then* will have been that we were going to have it all later, so that that must be among the things that we have now. Or formally:

C	(1)	Wp	
C	(2)	PWp	
K	(3)	$HFWp$	(1, $CpHFp$)
K	(4)	$PKWpFWp$	(1, 3, $CHpCPqPKqp$)
K	(5)	$PLCWpFWp$	(4, $CKWpqLCpq$)
K	(6)	$LCWpFWp$	(5, $PL = L$)
	(7)	FWp	(1, 6).

Again, if we have had it all *once* before, we have had it twice, $CWpCPWpPKWpPWp$. (Analogous reasoning.) And it can be shown metalogically that we can prove any theorem to the effect that if we have had it all at least n times before we have had it all at least $n+1$ times before; so that if we have had it all once before there is no limit to the number of times we have had it all before.

It might, in fact, be thought provable that if it is the case now that we have had it all before, then it has always been the case that we have had it all before, $CWpCPWpHPWp$. This does not, however, seem to be provable in any tense-logic of the kind we have been considering which leaves open the possibility (or positively asserts) that time is dense. The difficulty is to show that when we have an indefinite *number* of repetitions going back into the past, they must take us indefinitely *far* back; it could be (so far as we can show with this apparatus) that there is a point at which and before which we *don't* strike this 'world' again, though it is repeated an indefinite number of times as we approach this point. In a *metric* tense-logic such as we shall be considering in the next chapter, with variables for intervals, we can distinguish between being at different intervals from our imagined border, so that the worlds in this approaching series could *not* be *completely* the same if there were any such border; and in fact in such a logic *CWpCPWpHPWp* is provable. It is also provable in the logic of discrete time, which has, for instance, the thesis *CKPpPNPpPKpNPp* (allied to the thesis *CKFNpFGpFKNpGp* discussed in the last chapter). For let us suppose we have *Wp* and *PWp* but not *HPWp*, i.e. that we have *Wp*, *PWp* and *PNPWp* (*NH* = *PN*). We can then prove a contradiction thus:

	C	(1)	*Wp*	
*	*C*	(2)	*PWp*	
	C	(3)	*PNPWp*	
	K	(4)	*PKWpNPWp*	(2, 3, *CKPpPNPpPKpNPp*)
	K	(5)	*PLCWpNPWp*	(4, RPC)
	K	(6)	*LCWpNPWp*	(*PL* = *L*)
*		(7)	*NPWp*	(1, 6).

We can also prove, even without assuming discreteness or using interval-variables, that if we have had it all before, then we have also had-before everything between this time and the last time, i.e. *CWpCPKqPWpPKqPq*.

5. *The definition of tenses in terms of Diodorean modalities.* A further facet of the logic of total momentary states is the following: Diodorus defined the possible as what is or will be; is there any way of defining simple futurity in terms of Diodorean 'possibility'? As a start, one might try defining 'it will be' (*Fp*) as 'It is not, but either-is-or-will-be' (*KNpMp*, Diodorean *M*). But this obviously will not do; for 'it will be' is not understood as *excluding* 'it is', even though it does not entail it. P. T. Geach, however, has suggested a modification of this which is not open to this objection. 'It will be that *p*', he suggests, can be equated with 'For some *q*, *q* is not the case, but it either-is-or-will-be that both-*p*-and-*q*', i.e. *Fp* = *ΣqKNqMKpq*. For if *p* is going to be true later (whether it is true now or not), there will surely be *some* proposition which *will* be true contemporaneously with it but is *not* true now.[1] If we do define *Fp* in this way, and use

[1] Geach's definition was suggested by McTaggart's dictum that 'there could be no time if nothing changed' (*The Nature of Existence*, ch. xxxiii, § 309).

the system S4.3 for *M*, it is not hard to prove that *Mp* is equivalent to *ApFp*, in this sense of *F*; i.e. we can prove *EMpApΣqKNqMKpq* in quantified S4.3. Indeed, we can prove it in quantified M or T. It is equivalent to the following three implications.

1. *CMpApΣqKNqMKpq*
2. *CpMp*
3. *CΣqKNqMKpqMp*

(2 and 3 together) = *CApΣqKNqMKpqMp*.

Of these, 2 is an axiom of M. And since *A* = *CN*, 1 = *CMpCNpΣqKNqMKpq*, which follows by instantiation from *CMpCNpKNpMKpp* (*CMpMKpp* is of course in M and T). 3 = *ΠqCKNqMKpqMp*, which simply adds an antecedent in *CMKpqMp*. In other words we can prove in Geach's system the equivalence corresponding to the definition of the Diodorean *M* (as *ApFp*) in the ordinary tense-logical systems.

But can we prove in the ordinary systems, enriched with propositional quantifiers (with the ordinary rules for these), the equivalence corresponding to the definition of *F* in Geach's system, i.e. *EFpΣqKNqMKpq*, i.e. *EFpΣqKNqAKpqFKpq*? Here *KNqAKpqFKpq* is equivalent to *AKNqKpqKNqFKpq*, and since the alternative *KNqKpq* is self-contradictory it can be ignored, and what we have to prove is simply

EFpΣqKNqFKpq

Of the two implications that go to make up this, the proof of *CΣqKNqFKpqFp* is simple. This = *ΠqCKNqFKpqFp*, which follows from *CFKpqFp* (which we have even in Lemmon's minimal system K_t), but the converse implication *CFpΣqKNqFKpq* is another matter. Note, in the first place, that *ΣqKNqFKpq*, 'Some proposition is now false but is going to be true along with *p*', is equivalent to *ΣqKqFKpNq*, 'Some proposition'—namely the negation of the one we first thought of—'is now *true* but is going to be *false* at some time when *p* is true'. And this is the negation of *ΠqCqNFKpNq*, or *ΠqCqGCpq*. And this means that if we *could* prove that *Fp* implies *ΣqKNqFKpq*, we could prove it inconsistent with *ΠqCqGCpq*. But if *Fp* is inconsistent with this, it is inconsistent with the stronger proposition *ΠqCqHGCpq*, i.e. *Wp*. In other words, we could prove that if anything, say *p*, is going to be the case in the future, then this *p* cannot give the total present state of the world—that the present world-state is not going to be repeated. It is not really surprising that the equivalence corresponding to Geach's definition should give us this; for if p were the present total world-state, and Fp asserted its future repetition, there couldn't be a q that was false now but would be true at the time of p's later occurrence, for then that wouldn't be a genuine repetition of the *total* present state (which includes q's falsehood).

The consequences of Geach's suggestion can be developed with greater formal neatness if we put it in terms of G and the Diodorean L. Geach's definition of F yields a definition of Gp equivalent to $\Pi qCqLCNpq$, i.e. 'Any q which is now true, is and always will be implied by not p'—given Gp, this true q is (materially) implied by not-p *now* because it is true, and will be implied at all future times by not-p because at all future times not-p will be false. Alternatively, and equivalently, we might define Gp as $\Pi qCqLCNqp$, i.e. 'p is and always will be implied by the denial of any q which is now true'—given Gp, this denial will now imply p because it (i.e. Nq) is now false, and it will do so at all future times because p is then true and so 'implied' by *anything*. This last definition yields a neat proof of $CGCpqCGpGq$, using only the system T for L. We have to prove

$$C\Pi rCrLCNrCpqC\Pi sCsLCNsp\Pi tCtLCNtq,$$

and do it thus:

ΠtC	(1)	$\Pi rCrLCNrCpq$	
C	(2)	$\Pi sCsLCNsp$	
C	(3)	t	
K	(4)	$CtLCNtCpq$	(1, U.I.)
K	(5)	$CtLCNtp$	(2, U.I.)
K	(6)	$LCNtCpq$	(4, 3)
K	(7)	$LCNtp$	(5, 3)
	(8)	$LCNtq$	(7, 8)

($CLCpCqrCLCpqLCpr$ is in T). We also have

$$\vdash\alpha \rightarrow \vdash L\alpha$$
$$\rightarrow \vdash LCNqL\alpha \quad \text{(by } CLqLCpq\text{)}$$
$$\rightarrow \vdash CqLCNqL\alpha \quad \text{(by } CpCqp\text{)}$$
$$\rightarrow \vdash \Pi qCqLCNqL\alpha \quad \text{(by U.G.)}$$
$$\rightarrow \vdash G\alpha \quad \text{(by Df. } G\text{)}.$$

And we have

1.	*CLpp*	
2.	*CLpLCNqp*	(*CLqLCpq*)
3.	*CLpCqLCNqp*	(2, *CCpqCpCrq*)
4.	*CLpΠqCqLCNqp*	(3, *Π*2)
5.	*CLpGp*	(4, Df. *G*)
6.	*CLpKpGp*	(1, 5, *CCpqCCprCpKqr*)
7.	*CΠqCqLCNpqCpLCNpp*	(U.I.)
8.	*CGpCpLCNpp*	(7, Df. *G*)
9.	*CKpGpLCNpp*	(8, *CCqCprCKpqr*)
10.	*CKpGpLp*	(9, *CLCNppLp*)
11.	*ELpKpGp*	(6, 10),

which corresponds to the definition of the Diodorean *L* in ordinary tense-logic. But if, conversely, we could prove in ordinary tense-logic that *Gp* not only implies but is equivalent to *ΠqCqLCNpq*, this will make *GNp* equivalent to *ΠqCqLCpq*, which is implied by *Wp* (since what *permanently* implies every proposition now true, implies and always will imply it). Hence we would have *CWpGNp*, or *CWpNFp*, i.e. the present total truth will never be true again.

6. *Development of the U-calculus within the theory of world-states.* The 'worlds', or instantaneous total states of the world—the *p*'s such that *MWp*—of the present chapter, are clearly the same as the 'worlds' for which *a*, *b*, *c*, etc., may stand in the U-calculi sketched in Chapter III, and it is not difficult to bring these two 'logics of worlds' together. To do this, let us begin by slightly modifying both. Firstly, instead of treating the propositions of tense-logic, as they occur in the U-calculus, as *predicates* of worlds, and writing 'It is the case in the world *a* that *p*', or 'It is the case at the instant *a* that *p*', simply as *pa*, let us use the form *Tap*. Our basic stipulations then take the forms

U1. *ETaNpNTap*
U2. *ETaCpqCTapTaq*
U3. *ETaGpΠbCUabTbp*
U4. *ETaHpΠbCUbaTbp*.

Conditions on *U* may be stated as before, and proofs take very much the same shape. If only tensed propositions may be substituted for p and similar variables and if they enter the calculus only *as* substitutions for p, etc., then although it is not quite said that such propositions *are* predicates of worlds or instants, they do only occur as part of the form $Ta\alpha$ which *does* seem to predicate something of a world or instant, and which anyhow expresses a *function* of a world or instant, namely 'It is the case in (at)—that α'. But if the variables p, etc. are the very variables used in the propositional calculus to which, with quantification theory, the U-calculus is appended, then there would seem to be nothing syntactically wrong with such formulae as $TbTap$ or $TbUac$, or, conversely, with $FTap$ or $PUbc$. And there will be nothing *semantically* wrong with it either if the U-calculus can be given an interpretation *within* tense-logic. Such an interpretation, moreover, could be metalogically useful. It is easy enough to deduce tense-logical formulae preceded by Ta in the U-calculus, and to show, for example, that transitivity ($CUabCUbcUac$) gives the S4-type formula $TaCGpGGp$; but it would be good to have means also of showing that $CGpGGp$ gives transitivity.

Just such a translation is possible if we treat a, b, c, etc., as a sub-class of *propositional* variables, restricted to the (possible) world-*propositions* of the present chapter, for which we can lay down axiomatically

A1. Ma
A2. $ALCapLCaNp$

where $M\alpha = AA\alpha P\alpha F\alpha$ and $L\alpha = KK\alpha H\alpha G\alpha$. The variables a, b, c, etc., may be substituted for p, q, r, etc., in the basic tense-logic used (e.g. we have $CGCapCFaFp$ by substitution in $CGCpqCFpFq$), but not vice versa; for a, b, c, etc., we can only substitute other world-variables (e.g. we do not have $\vdash Mp$ from A1). Even complexes like Na are not substitutable for world-variables, though of course they are well-formed and are substitutable for p, q, r, etc. (If a expresses the total world-state at some instant, Na will not express the total world-state at any instant—unless there are only two instants—though of course it expresses *something*.) We may now define T and U as follows:

$$Tpq = LCpq$$
$$Upq = LCqPp\ (= TqPp).$$

These definitions are quite general; but in practice we consider mainly the special cases which have the forms *Tap* and *Uab*. *Tap*, 'It is the case in world *a* that *p*', is thus equated with *LCap*, 'The total world-state *a* is one which permanently implies that *p*', and *Uab*, 'World *a* is earlier than world *b*', with *TbPa*, 'It is true in world *b* that it has been the case that the world-state is *a*'. The equivalence (*EUabTbPa*) corresponding to this definition of *Uab* is provable in the U-calculus, if we add to it, for the truth in a world of a proposition which is itself a world, the stipulation *Taa* (every world is true in itself) and *CTabIab* (the *only* world-proposition which is true in any world is that world-proposition itself). We then have, for *CUabTbPa*,

C (1)	*Uab*	
K (2)	*KUabTaa*	(1, *Taa*)
K (3)	*ΣcKUcbTca*	(2, E.G.)
(4)	*TbPa*	(3, E (4) (3) from U4);

and for *CTbPaUab*, i.e. *CΣcKUcbTcaUab*,

ΠcC (1)	*KUcbTca*	
K (2)	*KUcbIca*	(1, *CTabIab*)
(3)	*Uab*	(2, *CIpqCϕpϕq*).

Our present concern, however, is not with proofs *within* the U-calculus, but with proofs *of* the postulates of the U-calculus within tense-logic supplemented by A1, A2, Df. *T*, and Df. *U*.

Positively, we can prove U1 and U2 from Lemmon's minimal tense-logic K_t with these supplementations. Splitting the equivalences U1 and U2 into their component implications, we have to prove

U1.1. *CTaNpNTap* U1.2. *CNTapTaNp*
U2.1. *CTaCpqCTapTaq* U2.2. *CCTapTaqTaCpq*.

We may begin with U2.1, which expands to *CLCaCpqCLCapLCaq*, this being provable in K_t as follows (the *L* fragment of K_t, it will be remembered, has all the laws of the 'Brouwersche' modal system, i.e. T+*CpLMp*):

1. *CCaCpqCCapCaq* (p.c.)
2. *CLCaCpqLCCapCaq* (1, RLC)
3. *CLCaCpqCLCapLCaq* (2, *CLCpqCLpLq*, Syll).

Of the others, U1.2 expands to *CNLCapLCaNp*, which, since $A = CN$, is just A2. U1.1 expands to *CLCaNpNLCap*, which amounts to the denial of the conjunction *KLCaNpLCap*. We show this conjunction false (and so U1.1 true) by proving from it the denial of *Ma*, i.e. A1, thus:

C	(1)	*LCaNp*	
C	(2)	*LCap*	
K	(3)	*LCaKpNp*	(2, 1, *CLCpqCLCprLCpKqr*)
K	(4)	*LNa*	(3; *NKpNp*, RL; *CLCpqCLNqLNp*)
	(5)	*NMa*	(4, $LN = NM$).

(The theses appealed to in the proofs of lines (3) and (4) may be proved in the same way as U2.1.) Finally we prove U2.2 from the rest without expanding *T*, except in this proof of a rule (call it RT) to infer $\vdash T a\alpha$ from $\vdash \alpha$:

$$\vdash \alpha \rightarrow \vdash Ca\alpha \quad \text{(by } CpCqp\text{)}$$
$$\rightarrow \vdash LCa\alpha \quad \text{(by RL).}$$

We then prove *CCTapTaqTaCpq* by similar steps to those used in the proof of *CCTpTqTCpq* in Scott's 1964 system for discrete time (Section 3 of Chapter IV).

But whether U3 and U4 are provable without a strengthening of the basis somewhere, is less certain. The best I have been able to come up with are the following 'proofs' of *CTaGpΠbCUabTbp* and *CTaHpΠbCUbaTbp* (which are implicational halves of U3 and U4):

ΠbC	(1)	*LCaGp*	(= *TaGp*)
C	(2)	*LCbPa*	(= *Uab*)
K	(3)	*LHCaGp*	(1, ?)
K	(4)	*LCPaPGp*	(3, *CHCpqCPpPq*)
K	(5)	*LCPap*	(4, *CPGpp*)
K	(6)	*LCbp*	(2, 5, *L*-syll)
	(7)	*Tbp*	(6, Df. *T*).

For our half of U4, we can prove the lemma *CKbpLCbp*, thus:

C	(1)	*Kbp*	
K	(2)	*NCbNp*	(1, p.c.)
K	(3)	*NLCbNp*	(2, *CLpp*)
	(4)	*LCbp*	(3, A2),

and then we have, for the main theorem:

ΠbC	(1)	*LCaHp*	(= *TaHp*)
C	(2)	*LCaPb*	(= *Uba*)
K	(3)	*LCaPKbp*	(1, 2, *CKHpPqPKpq*)
K	(4)	*LCaPLCbp*	(3, lemma)
K	(5)	*MPLCbp*	(4, A1, *CLCpqCMpMq*)
K	(6)	*LCbp*	(5, ?)
	(7)	*Tbp*	(6, Df. *T*).

We can remove the queries from these proofs if we can prove *CLCapLHCap* and *CMPLCapLCap* in K_t+A1+A2. And if we *can* do this, the following result, in which we add *CPPpPp* to our basis and prove the transitivity of *U*, is significant:

C	(1)	*LCbPa*	(= *Uab*)
C	(2)	*LCcPb*	(= *Ubc*)
K	(3)	*LHCbPq*	(1, *CLCapLHCap* ?)
K	(4)	*LCPbPPa*	(3; *CHCpqCPpPq*, RL)
K	(5)	*LCPbPa*	(4, *CPPpPp*)
K	(6)	*LCcPa*	(1, 2, 5)
	(7)	*Uac*	(6, Df. *U*).

In this line of investigation, as in others, we can probably dispense with world-variables, if we wish, by adding to our theses conditions corresponding to the axioms A1 and A2, e.g. in the calculus without world-variables we would aim to prove, instead of *ETaNpNTap*, the thesis

CKMpΠqALCpqCLpNqETpNrNTpr.

7. *'States' consisting of combinations of Hamblin tenses.* In the type of tense-logic for which Hamblin's 15-tense theorem holds, there is a species of 'state' proposition which is not at all a complete world-state proposition, but which nevertheless has some logical interest. This is a conjunction with fifteen conjuncts, each of which is one of the fifteen tenses or its negation, each of the tenses being covered one way or the other, and all applied to a single proposition. There are 2^{15} different conjunctions, each incompatible with each of the others. Indeed the great majority of them are internally inconsistent; but there are upwards of 50 which are not. In all of these there are redundant components

which can be sheared off, since if one of the fifteen tenses is affirmed all those which it implies can be taken for granted, and if one of them is denied, the denials of all those which imply it can be taken for granted. Some specimens are the following:

(*a*) *HGp*; this implies the affirmation of all the rest.
(*b*) *KKPGpNHpPHp* (= *KKPGpPNpPHp*) where *PGp* implies *Gp*, *FGp*, *GFp*, *Fp*, *HFp*, *PFp* *p*, and *GPp*; *PHp* implies *HPp*; and *NHp* implies *NFHp* and *NHGp*.
(*c*) *KPGpNHPp* (= *KPGpPHNp*).

We might diagram these three as follows, with the vertical line for the present moment, past to the left, future to the right, an open strip above the horizontal for times of truth, and a filled in one below it for times of falsehood:

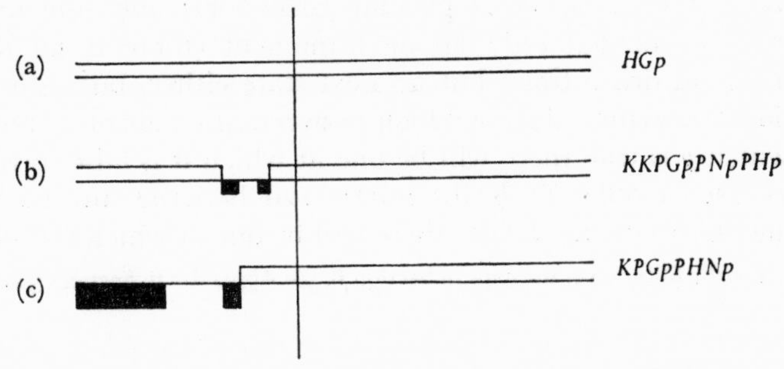

The bits without covering or shading may be filled in in different ways which propositions of the class we are considering do not distinguish (they do give a set of mutually exclusive and collectively exhaustive determinations of *p*, but finer discriminations are possible within most of them). One thing which is clear with each of these three, however, is that they have come to stay. In the diagram, moving the vertical line to the right makes no difference to the general shape of the picture, and formally, for each state α in this group (we may call them 'Kribs states', after the initiator of this line of investigation), we may prove $C\alpha G\alpha$. The proofs are quite simple; that for (*c*) goes thus:

1. *CPGpGPGp* (*CPpGPp*, *p*/*Gp*)
2. *CPHNpGPHNp* (ibid., *p*/*HNp*)

3. *CKPGpPHNpKGPGpGPHNp* (1, 2, *CCprCCqsCKpqKrs*)
4. *CKPGpPHNpGKPGpPHNp* (4, *CKGpGqGKpq*).

Not all Kribs states are thus permanent; some of them, indeed, are essentially borderlines, and cannot have any duration at all. This applies, for instance, to all the states which combine the denial of *PGp* or of *FHp* with the affirmation of *Gp* or of *Hp*. The simplest case of this sort is *KKNpHpGp*, '*p* is false now but always has been true and always will be'. And most of the states are bound, by the information contained within them, to give place to different ones sooner or later, and to go through a cycle which finishes either with something permanent (some combination either of *Gp* or of *GNp* with various additions about the past) or with an oscillating pair (containing or implying *KGFpNFGp*, i.e. *KGFpGFNp*, or *GKFpFNp*). Sometimes what the next state will be is unambiguous, and sometimes there will be alternative possible successors; and sometimes there may be not only no next moment (there is no next moment in dense time) but no next state either, but rather a period of fuzziness during which between any pair of moments in which *p* is true there will be one at which it is false (giving, of course, a different Kribs state), and between any pair of moments at which *p* is false there will be one at which it is true.

Note. For the filling-in of the gaps in Section 6, see Appendix B, Sections 3 and 4.

VI

METRIC TENSE-LOGIC

1. *The syntax of intervals.* I have mentioned the possibility of enriching tense-logic with variables representing intervals. A system of this sort was sketched in *Time and Modality*,[1] with the form *Pnp* for 'It was the case the interval *n* ago that *p*', and *Fnp* for 'It will be the case the interval *n* hence that *p*'. Along with these go quantifiers, giving us *ΣnFnp* for 'For some *n*, it will be the case the interval *n* hence that *p*'; *ΠnFnp* for 'For all *n*, it will be the case the interval *n* hence that *p*'; and similarly with *P*. *ΣnFn*, *ΠnFn*, *ΣnPn*, *ΠnPn* may be respectively abridged to the *F*, *G*, *P*, and *H* of the preceding chapter, provided that there is no free *n* in what follows them.

The proviso is necessary, because, for example

(A) 'For some *n*, it will be the case the interval *n* hence that both (i) I am drinking and (ii) it will be the case the interval *n* later that I am ill' (*ΣnFnKpFnq*)

means something slightly different from

(B) 'It will be the case sooner or later that both (i) I am drinking and (ii) it will be the case the interval *n* later that I am ill' (*FKpFnq*).

For (A) is a complete proposition, and means that some time after now I shall be drinking, and exactly the same amount of time after that I shall be ill. (B), on the other hand, is a still open sentence, and doesn't say anything definite until the variable *n* is replaced by a specific interval or else bound by a new quantifier somewhere. If we put a *Σn* at the beginning of (A) it will be vacuous, and leave the sense unaltered; if we put it at the beginning of (B) it will give us

(C) For some *n*, it will be the case sooner or later that both (i) I am drinking, and (ii) it will be the case the interval *n* later that I am ill',

[1] Ch. ii.

and this means something a little less specific than (A), namely that I am going to be drinking and then be ill, without anything about the illness being twice as far away from now as the drinking. The *F* of (B) and (C) can, however, be replaced by a quantification over intervals, provided that the variable used is not *n*, e.g. we could give (C) as $\Sigma n \Sigma m FmKpFnq$.

It ought not to be necessary to say that quantifications of this sort do not imply that intervals are entities. $\Sigma nPnp$, 'It was the case at some time or other that *p*' is just a generalization of remarks like 'It was the case this time yesterday that *p*', in which there are no named entities except any which may be named by expressions within *p*. There is, however, a more subtle mistake that may be made here. In this symbolism the *n* has no meaning apart from the preceding *P*, and it cannot get into the proposition that follows it except in the company of that *P*. In ordinary speech we can be misled into carving sentences up in a different way. 'I was sick yesterday' suggests that 'yesterday' modifies 'sick', and that being sick-yesterday is a particular way of being sick. It is not; it is, if it is anything at all like that, a way of *having been sick*; and more accurately, 'having been yesterday' is a way of having been. Buridan[1] has an instructive puzzle about this. If Socrates will run tomorrow (*Sortes curret cras*), is it true to say that he will be running tomorrow (*Sortes erit currens cras*)? The problem here has nothing to do with differences between performances and activities; put all that on one side. The *pro* argument is that 'Socrates will be running tomorrow' is the normal and proper way of putting 'Socrates will run tomorrow' into the standard logical form with subject (*Sortes*), copula (*erit*) and predicate (*currens*). The difficulty is one of applying the rule that a future-tense proposition *is* true if and only if the corresponding present-tense proposition *will be* true. For 'Socrates *runs* tomorrow', i.e. *Sortes currit cras* with the ordinary present tense (not the journalistic 'runs'), is never true and never will be. Buridan's answer is that when the verb is spread out into copula and predicate, 'tomorrow' does not modify the predicate but the copula; the predicate is not *currens cras*, but the copula is *erit cras*. And when the rule for the truth of such propositions is being applied, the 'tomorrow' must be taken right out of the present-tense pro-

[1] *Sophismata*, ch. 4, *sophisma* 5.

position which is going to be true then; it belongs in fact with the 'will be true' that is said of it—the present-tense *Sortes currit* (with the 'tomorrow' out of it) will-tomorrow be true. Similarly with 'Socrates argued last year'; it should be spread out to 'Socrates was-last-year arguing', and it's plain Socrates-is-arguing which was-last-year the case.

2. *Postulates for metric tense-logic.* In axiomatizing the metric system, it is convenient to think of variables m, n, etc., as representing numbers measuring the intervals. These may be drawn from the real numbers, or only from the rationals, or only from the integers; which we do here will influence what is provable from our postulates. Other differences depend on whether we draw upon the whole range of such numbers—positive, negative, and zero; or only upon those greater than or equal to zero, i.e. zero and the positives; or only upon the positives. In theses we may substitute for them any expressions which denote numbers of the sort being used, and we may replace any such expression by an arithmetically equivalent one, e.g. m by $(n+m-n)$.

If we use the whole range, only one tense-logical primitive will be needed; let it be F, and let Pnp be defined as $F(-n)p$. We may then subjoin to propositional calculus and quantification theory the rule

$$\text{RF: } \vdash \alpha \longrightarrow \vdash Fn\alpha$$

and the axioms

FO : $CFOpp$	FC : $CFnCpqCFnpFnq$
FN1 : $CFnNpNFnp$	FF : $CFmFnpF(m+n)p$
FN2 : $CNFnpFnNp$	FΠ : $C\Pi nFmFnpFm\Pi nFnp$.

These (apart from the last, which replaces an FΣ) are substantially the postulates in *Time and Modality*, except that there they are set up for the future only (apart from FO), with negative values of n excluded. Similar postulates are used by Rescher for the calculus with negative numbers allowed, and with Pn as $F(-n)$;[1] he uses $C\Pi nFnpp$ in place of FO, and the law

$$\text{FK: } CFnKpqKFnpFnq$$

[1] Nicholas Rescher, 'The Logic of Chronological Propositions', *Mind*, Jan. 1966.

in place of FC. These last differences are of course trivial; it is, as Rescher points out, an easy matter to prove either set of postulates from the other. He also points out that, given FN1 and FN2, it is easy to prove the converses of the remaining axioms.

The use of negative numbers, however, is far from trivial. It enables Rescher and Meyer to prove, for example, $CFm\Pi nFnp\Pi nFnp$, from which we get $C\Sigma mFm\Pi nFnp\Pi nFnp$, i.e. $CFGpGp$. In the proper sense of 'It will be that' and 'It will always be that', this is quite counter-intuitive, but in Rescher's system F in effect means 'It is or has been or will be that' and G 'It is and always has been and always will be that', which gives them the properties of the M and L of S5 (e.g. $CMLpLp$). In this system it doesn't in fact matter whether we choose our measure numbers from the reals, the rationals, or the integers; the differences between discrete, merely dense and continuous time do not appear—as far as the symbols go, the system has *all* the laws, trivially (e.g. it has $CFpFFp$ because it has $CpFp$, and it has $CKFNpFGpFKNpGp$ because both antecedent and consequent are inconsistent—think of them as $KMNpMLp$ and $MKNpLp$ in S5, where $MLp = Lp$).

If we want to make finer distinctions within the system, we must reinstate the difference between past and future. As a first step, we may exclude negative numbers from our interval-measures (still leaving zero), replace the definition of Pnp as $F(-n)p$ by a mirror-image rule, and add the mixing axioms

FP1 : $CFmPnpF(m-n)p$, for $m \geqslant n$
FP2 : $CFmPnpP(n-m)p$, for $n \geqslant m$
FPΠ: $C\Pi nFmPnpFm\Pi nPnp$.

The provisos on FP1 and FP2 are of course needed because only non-negative numbers are to be used in the formulae. We add, however, a rule that if something holds under both provisos we may drop them. The following proof will illustrate this:

$\Pi m\Pi nC$ (1) Fmp
C (2) Fnq
K (3) $F(m+n-m)q$ (2)
K (4) $FmF(n-m)q$, for $n \geqslant m$ (3, Cnv. FF)
K (5) $FmKpF(n-m)q$, for $n \geqslant m$ (1, 4, Cnv. FK)

K (6) $FmKp\Sigma lFlq$, for $n \geqslant m$ (5, EG)
K (7) $AFmKp\Sigma lFlqFnKq\Sigma kFkp$, for $n \geqslant m$ (6, $CpApq$)
K (8) $F(n+m-n)p$ (1)
K (9) $FnF(m-n)p$, for $m \geqslant n$ (8)
K (10) $FnKqF(m-n)p$, for $m \geqslant n$ (2, 9)
K (11) $FnKq\Sigma kFkp$, for $m \geqslant n$ (10)
K (12) $AFmKp\Sigma lFlqFnKq\Sigma kFkp$, for $m \geqslant n$ (11, $CpAqp$)
K (13) $AFmKp\Sigma lFlqFnKq\Sigma kFkp$ (7, 12, drop provisos)
K (14) $A\Sigma mFmKp\Sigma lFlq\Sigma nFnKq\Sigma kFkp$ (13, EG)
(15) $AFKpFqFKqFp$ (14, Df. F).

What has been proved here is in effect $CKFpFqAFKpFqFKqFp$, i.e. Hintikka's axiom for S4.3 with F for M. Since Fnp in this system includes the case FOp, a 'zero future' which is equated with the present, the proper meaning of the form $\Sigma nFnp$ is 'It *is or* will be the case that p', i.e. Diodorean possibility. In the full past-future calculus developed in this way, we have a basis for Hamblin's original system, with F and P for Diodorean possibility and its mirror image; at least, we have exactly that if we let our measure-numbers be the rationals, and at least that if we let them be the reals or the integers.

It is important to notice that in the above proof we are *not* able to proceed as follows:

—K (5) $FmKpF(n-m)q$, for $n \geqslant m$
—K (6′) $AFmKpF(n-m)qFnKqF(m-n)p$, for $n \geqslant m$ (5, $CpApq$)
—K (10) $FnKqF(n-m)p$, for $m \geqslant n$
—K (11′) $AFmKpF(n-m)qFnKqF(m-n)p$, for $m \geqslant n$ (10, $CpAqp$)
—K (12′) $AFmKpF(n-m)qFnKqF(m-n)p$ (6′, 11′, drop provisos)

For (6′), (11′), and (12′), except where $m = n$, are ill-formed on *both* provisos (in each case, unless $m = n$, either $n-m$ or $m-n$ will be a minus quantity).

If we distinguish the present totally from the past and the future by discarding the forms FOp and POp and drawing upon positive numbers only for our interval-measures, we must change the provisos in FP1 and FP2 above to 'for $m > n$' and 'for $n > m$', and add the further axiom

$$FP3 : CFnPnpp.$$

This amounts to '$CFmPnpp$, for $m = n$', and provisos are now to be dropped when something holds under all three of the possible ones. Within this basis we can construct *GH*-calculi or *PF*-calculi of the more standard sort described in Chapter III. The above proof of Hintikka's law for the Diodorean *M*, for example, can be replaced by a similar proof (using three provisos instead of two) of the analogous law, with three alternatives, for the proper future, i.e.

$$CKFpFqAAFKpqFKpFqFKqFp.$$ [1]

The device of provisos can be dropped, at least in the middle system, if we incorporate not only arithmetical expressions but also arithmetical propositions, such as $m \geqslant n$, into the body of our calculus; but this must be done with circumspection. It will not do, for example, to replace FP1 (in the third system) by

$$C(m \geqslant n)CFmPnpF(m-n)p,$$

for when $m < n$, $F(m-n)p$ will still be ill-formed. Geach, however, has pointed out that this difficulty may be overcome by using $|m-n|$ for the *absolute* difference between m and n, i.e. the non-negative number that measures the difference between them, whichever way it goes. FP1 and FP2 can then be replaced by the pair:

$$C(m \geqslant n)CFmPnpF|m-n|p$$
$$C(n \geqslant m)CFmPnpP|m-n|p.$$

As with the non-metric systems, if we drop the mirror-image rule in favour of separate mirror images of the other rules and axioms, it is not necessary to do this with all of them. For example, given the rest, at least PP is derivable from FF, and in the 'middle' system, PO from FO, thus:

$$Pop \rightarrow PoFop \text{ (PF1, conv.)} \rightarrow Fop \text{ (PF2)} \rightarrow p \text{(FO)}.$$

[1] For another proof within this calculus, and for further discussion of metric tense-logic generally see A. N. Prior, 'Postulates for Tense Logic', *American Philosophical Quarterly*, April 1966.

3. *Interaction of the A series and the B series.* This apparatus enables us to state precisely some relations between McTaggart's A series and his B series. It is important, as we have seen, not to treat the A series as if it were a B series; just that constitutes McTaggart's Fallacy. It was, however, practically his *only* fallacy in this area, and it should not lead us to imagine that the A series and the B series are so distinct that they cannot be brought into a common context. As McTaggart said, the A series 'slides along' the B series and vice versa; 'later and later terms pass into the present' and 'presentness passes to later and later terms'. It is a particular merit of Rescher's article, referred to above, that he makes it quite clear that what he calls 'chronologically indefinite' time-references can occur within what he calls 'chronologically definite' ones, i.e. that what can be or not be the case *at* a given date may be something tensed, e.g. that it will be raining, and the relations between the two series can be given by such simple rules as that it is (permanently, or maybe even tenselessly) the case at t that 'it will be the case the interval n hence that p' if and only if it 'is' the case at $t+n$ that it simply is (present-tense) the case that p.

It is also clear from Rescher's article that we may, conversely, embed dated propositions within tensed ones. It has, indeed, been pointed out by Broad[1] that our ordinary use of dates and of words like 'earlier' and 'later' is tensed rather than tenseless. 'Before either battle had happened it would have been true to say "There will be a battle at Hastings and there will be a battle at Waterloo 749 years later". . . . During the battle of Hastings it would have been true to say "There is a battle going on at Hastings and there will be a battle at Waterloo 749 years later". At any intermediate date it would have been true to say', etc., etc. 'No one but a philosopher would say, "The Battle of Hastings precedes the Battle of Waterloo by 749 years".' Moore also, commenting on McTaggart's 'If M is ever earlier than N, it is always earlier', similarly comments, 'Queen Anne's death *was* earlier than Marlborough's (merely another way of saying "Anne died before Marlborough"): that is true *now*; but it was not always true; e.g. in 55 B.C. it was true that Anne *would* die before Marlborough, but not that she *did* die before Marlborough. The only thing

[1] *Examination of McTaggart's Philosophy*, vol. 2, pp. 29–89.

that both *was* true in 55 B.C. and *is* true now is the proposition "Either Anne *will* die before Marlborough, or Anne *did* die before Marlborough, or Anne *is* dying before Marlborough", and this *alternative* proposition certainly *is* true now, *was* true at every past moment, and *will be* true at every future.' Dated propositions, in short, are not a-temporal, but certain disjunctions of dated propositions are true at all times. McTaggart apparently 'imagines that "either was at some time earlier, or is now earlier, . . . or will at some time be earlier" entails some proposition that could be expressed by "is earlier" where "is" is used *timelessly*, as it is said to be in "twice 2 *is* 4": but is there any such proposition? If there isn't, then he is using it as short for the disjunction.'[1]

Even the 'omni-temporality' of such disjunctions, and of forms that may be 'short for' them, means that the prefixing of tense-operators to them (with or without stated intervals) is a little trivial. The rule of truth for such complexes would be simply that 'It is (was, will be) the case that p', where p is of this sort, is true if and only if the simple p is. If, however, it is maintained that either dated propositions or any other propositions (e.g. Moore's example of '$2+2 = 4$') are non-temporal in the sense that it 'makes no sense' to prefix tense-operators to them, we do encounter one serious problem, namely, does it make sense to prefix such operators to compounds, e.g. conjunctions and disjunctions, of which one part is temporal and the other not? Wittgenstein says that 'the logical product', i.e. conjunction, 'of a tautology and a proposition says the same as the proposition. Therefore that product is identical with the proposition.'[2] Equating non-temporal propositions with Wittgenstein's 'tautologies', if they are true, and with his 'contradictions', if they are false, this would suggest that if we use a, b, etc., for non-temporals and p, q, etc. for temporals, Kap is the same proposition as p when a is true, and the same as a when that is false; and Aap is the same as p when a is false, and as a when that is true. It is certainly the case that if a is timelessly true, the truth-value of Kap will be liable to vary with

[1] *The Commonplace Book of G. E. Moore*, pp. 404–5. Cf. also Moore's *Lectures on Philosophy*, pp. 9–10.

[2] *Tractatus*, 4. 465. The possible relevance of this passage to this problem, and the solution it suggests, were pointed out to me by Miss G. E. M. Anscombe.

that of *p*, while that of *Aap* will be timelessly fixed as true, and if *a* is timelessly false, the truth-value of *Aap* will be liable to vary with that of *p*, while that of *Kap* will be timelessly fixed as false. But if this means that it makes sense to prefix, say, 'It will be the case that' to *Kap* if *a* is true and does not if it is false, and that the converse holds with *Aap*, this is a very awkward formation-rule indeed. Still, in a formal calculus one could perhaps allow 'vacuous' tense-operators to be prefixed to non-temporal propositions. Or one may question, with Moore, whether there are in fact any such. '"5 is a bigger number than 3" is said to be "*timelessly*" true; but we certainly can correctly say "is bigger now, always *was* bigger, and always *will be* bigger".'[1]

4. *The logic of dates.* The interactions between the A series and the B series which emerge from Rescher's paper may be summed up as follows: If we use the form *Tap* to mean 'It is the case at the date *a* that *p*', its laws are very similar to those of the *Fnp* of our simplest interval-calculus, the one in which *Pnp* is defined as $F(-n)p$. They are as follows:

$$\text{RT}: \vdash\alpha \rightarrow \vdash Ta\alpha$$

$$\text{TN1}: CTaNpNTap \qquad \text{T}\Pi\text{1}: C\Pi aTapp$$

$$\text{TN2}: CNTapTaNp \qquad \text{T}\Pi\text{2}: C\Pi bTbpTa\Pi bTbp$$

$$\text{TC}: CTaCpqCTnpTaq \qquad \text{TT}: CTaTbpTbp$$

The difference appears in the last axiom where the corresponding law FF is $CFmFnpF(m+n)p$. If we co-ordinate the numbers used in dating with those used in interval-measurement, we also have $CTaFmpT(a+m)p$, and (since dated statements do not alter in truth-value) *CFmTapTap*. If we use the *F* and *P* of our third interval-calculus, in which interval-variables stand for positive numbers, but allow date-variables to stand for positive, negative, and zero numbers, the mixing laws are

$$\text{TF}: CTaFnpT(a+n)p$$
$$\text{TP}: CTaPnpT(a-n)p$$
$$\text{FT}: CFnTapTap$$
$$\text{PT}: CPnTapTap.$$

(The converses of these are derivable.)

[1] *Commonplace Book*, p. 405. Cf. also *Some Main Problems of Philosophy*, p. 294: 'For my part, I cannot think of any instance of a thing, with regard to which it seems quite certain that it *is*, and yet also that it is not *now*.'

These mixing laws, together with the pure T postulates, may be obtained within our third interval calculus (with a slight enrichment) if we define dates in terms of intervals in substantially the way in which this is done in actual dating systems. To say that a certain event occurred in A.D. 1066 is to say, approximately, that it occurred 1066 years after the birth of Christ (or the putative birth of Christ, i.e. after a time so many years before the Church gave the calendar its present shape). That is, it is to say that it was the case when the event occurred that it had been the case 1066 years before that Christ was being born; or to use our tidied-up version of English, that it was the case then that both (*a*) the event is occurring, and (*b*) it was the case 1066 years ago that Christ is being born. Formally, we introduce into metric tense-logic a propositional constant ϕ, representing the event taken as the origin of our dating system; and using $M\alpha$ as short for $AA\alpha P\alpha F\alpha$ (cf. Moore's interpretation of McTaggart's 'is'), we give the following three-part definition of the form Tap:

$$Tap = MK\phi p, \text{ for } a = 0$$
$$Tap = MKPa\phi p \text{ (or } MK\phi Fap\text{), for } a > 0$$
$$Tap = MKF(-a)\phi p \text{ (or } MK\phi P(-a)p\text{), for } a < 0.$$

For example, 'It is or has been or will be the case that (p at the date -144.6)' is translated as 'It is or has been or will be that both (*a*) it will be the case 144.6 years hence that Christ is being born, and (*b*) p', or as 'It is or has been or will be that both (*a*) Christ is being born, and (*b*) it was the case 144.6 years ago that p'.[1]

Some of the postulates above, e.g. TF and TP, follow very easily from these definitions and the postulates earlier laid down for metric tense-logic. Take TF, i.e. $CTaFnpT(a+n)p$. For $a = 0$, the antecedent $= MK\phi Fnp$, and the consequent $= T(0+n)p = Tnp$, with $n > 0$, which again $= MK\phi Fnp$. For $a > 0$, the antecedent

$$TaFnp = MK\phi FaFnp = MK\phi F(a+n)p,$$

and the consequent $T(a+n)p$, $a+n$ being greater than 0, also $= MK\phi F(a+n)p$. Where $a < 0$, either a is *numerically* greater than n, i.e. the *positive* number $(-a)$ is greater than n, or $(-a) < n$. In the first case, $TaFnp = MK\phi P(-a)Fnp$, which,

[1] Cf. *Time and Modality*, p. 19.

since $(-a) > n$, $= MK\phi P(-a-n)p = MK\phi P(-(a+n))p$. And in this case $T(a+n)p$, since $a+n < 0$, also $= MK\phi P(-(a+n))p$. In the other case, where $(-a) < n$ and $a+n$, or $n-(-a)$, > 0, $TaFnP = MK\phi P(-a)Fnp = MK\phi F(n-(-a))p = MK\phi F(a+n)p = T(a+n)p$.

Others of the Rescher postulates, e.g. TN2 (*CNTapTaNp*) require for their proof, in addition to the definitions and ordinary metric tense-logic, some special postulate or postulates for the constant ϕ, e.g. that it is a 'world' proposition in the sense of the last chapter, or the rather stronger postulate (entailing that time is not circular) that ϕ is true at a single instant only, $MK\phi NP\phi NF\phi$. In connexion with the second alternative, it is worth observing that an important theorem about unique propositions, i.e. ones true at a single instant only, is provable in metric tense-logic, namely that if at any time there is a proposition true at that time only, then at every time there is a proposition true at that time only. In particular, if p is true at one time only, *Pnp* and *Fnp* are true at one time only, for each n. So if ϕ, the origin-event, is true at one time only, then every statement of the form 'It was the case the interval n ago that ϕ' and 'It will be the case the interval n hence that ϕ' is true at one time only. This is a variant of the argument that no instantaneous world-state can ever be repeated in its totality, for at A.D. 1966.23, for instance, the totality of truth will include the truth that it is A.D. 1966.23, and this will not be true at any other time. We can now see that this argument can be put forward without assuming that there are actual objects called Dates which acquire Presentness and then instantly lose it; all that the argument need mean is that it is or has been or will be true once only that it was the case the interval n ago that the-origin-event-is-occurring (and true once only that this will be the case the interval n hence). The argument only works in this form, however, if uniqueness can be postulated for the origin-event itself.

5. *Metric circular time.* If time is taken to be circular, the introduction of specific intervals makes it possible to distinguish the two ways of going round the circle. In circular time (on its simplest interpretation), whatever will be has been, and whatever will always be has always been, and vice versa, so that

we cannot distinguish G from H or F from P; but it does not follow that what will be the case this time tomorrow was the case this time yesterday, i.e. we do *not* have $EPnpFnp$ as a law. What we do have in circular metric tense-logic is a new sort of constant, the interval κ that represents a complete cycle; and for this we have such laws as

(1) $EF\kappa pp$,

and where $s\kappa$ represents any integral multiple of κ that is greater than n,

(2) $EFnpP(s\kappa - n)p$,

e.g. if κ itself is greater than n, we have

(3) $EFnpP(\kappa - n)p$

From (2), $C\Pi nPnp\Pi nFnp$, i.e. $CHpGp$, is easily deducible, as $CGpp$ is from (1).

6. *Enlargement of tense-logic to make metric concepts definable.* It is arguable that not only the use of dates but the use of measured intervals is a comparatively sophisticated and artificial procedure, and measured intervals ought to be definable within a tense-logic of a more 'primeval' sort.[1] Just as dating in practice involves the introduction of an origin-event into a logic of measured intervals, so interval measurement involves the synchronizing of events with the phases of some cyclical process. 'It was the case this time yesterday that p', for example, amounts to 'It was the case that p when the sun was last in its present position'. The theory of interval measurement would therefore appear to be built upon propositions of the form 'It was the case that p the last time it was the case that q'. But propositions of this form do not seem to be definable in terms of the indefinite P and F of our first type of tense-logic. For example, $PKNqPKqp$ does not give us quite what we want; it means 'It was the case that p on *some* previous occurrence of q, separated from now by *at least* one moment or period of q's falsehood'. This is compatible with q and Nq having alternated

[1] The artificiality of quantification over intervals is stressed by P. T. Geach in his review of *Time and Modality* in the *Cambridge Review*, 4 May, 1957, p. 543.

more than once between now and the time we say that p and q were true together. On the other hand,

$$K(PKNqPKqp)(NPKqPKNqPKqp)$$

is too strong. For while this indeed says that p was true the last time that q was, it also says that p was never true on any occasion of q's truth previous to that one, and we want to leave that open.

Hans Kamp has pointed out (1966) that what is needed here is something in between the merely 'topological' tense-logic with P, H, F, and G and the fully 'metric' sort with Pn and Fn; and he has begun the development of just such a system. As primitives it uses a pair of two-place functions which may be represented as Φpq and Ψpq. The former of these means 'q has been true from some past time at which p was true, up to (but not necessarily including) now'. In a metric tense-logic supplemented by arithmetic, this function would be equivalent to

$$\Sigma nKPnp\Pi mC(m < n)Pmq,$$

and in a U-calculus with Uab interpreted as 'a before b' we would have

$$ETa\Phi pq\Sigma bKUbaKTbp\Pi cCKUbcUcaTcq,$$

i.e. Φpq is the case at a if and only if for some b earlier than a, p is true at b, and for all c between b and a, q is true at c. But these, at all events the first, are not definitions; this is a more fundamental calculus in terms of which there is some hope that Pnp can be defined. Ψ is simply the future-tense analogue of Φ. What certainly can be defined in terms of Φ is the desired function 'p the last time that q', which is $\Phi KpqNq$, 'q has been false from some past time at which p and q were true together, up to now'. We can also define the simple Pp as $\Phi pCpp$, 'The tautology Cpp has been true from some past time at which p was true, up to now'.

The converse function $\Phi Cppp$, which we may abridge to $H'p$, is of some interest also. It may be read as 'p just before now', meaning that p has been uninterruptedly true from some past time up to (but not necessarily including) now. This function is not equivalent to, though with dense time it implies, $NH'Np$, which it is therefore useful to abridge to $P'p$. Both

H'p and *H'Np* will be false if there is what I have called a 'fuzz' of *p*'s and *Np*'s in the immediate past, i.e. if between any past moment of *p*'s truth and the present, however close, there is a moment of *p*'s falsehood, and conversely. This counter-example is not available, of course, in discrete time, and indeed in discrete time *H'p* is true, in a vacuous way, whatever *p* might be. For *H'p* is true if *p* is true at all times between some past time and the present; but in discrete time there is one past time (the one *just* past) such that there is *no* time between it and the present, so that any proposition to the effect that *if* a time is between that one and the present, *p* is true at it, will be vacuously true. But in dense infinite time $Hp \rightarrow H'p \rightarrow P'p \rightarrow Pp$. Kamp has investigated the 'tenses' constituted by sequences of *P*, *H*, *F*, *G*, *P'*, *H'*, *F'*, and *G'*, and has found that although they are infinite in number (even in the dense infinite time that yields only fifteen for *P*, *H*, *F*, *G* on their own) they have quite a definite implicational pattern. In discrete time, Kamp has also pointed out, Scott's function *Yp*, '*p* at the moment just past', is definable as *Φpp*. (If *p* was true at the moment just past, there is a moment, *viz.* the one just past, at which we have both *p*-true-then, and also *p*-at-all-times-between-then-and-now, since there is *no* time between then and now. And if *p* was true at some past time and at all times since, it was true at the moment just past.)

For a start towards the axiomatization of this area of tense-logic, the *U*-counterparts of the following postulates for *Φ* and *Ψ* (Prior, 1966) are provable without imposing any conditions on the relation *U*, and they are sufficient, with their mirror images, to yield the whole of Lemmon's minimal system K_t:

R1: $\vdash\alpha \rightarrow \vdash N\Phi N\alpha\gamma$
R2: $\vdash C\alpha\beta \rightarrow \vdash C\Phi\gamma\alpha\Phi\gamma\beta$
A1: *CNΦNCpqrCΦprΦqr*
A2: *CΦNΨNpqqp*.

The key proofs are as follows:

T1: *CCppCqq* (p.c.)
T2: *CΦrCppΦrCqq* (T1, R2)
T3: *CNΦNCpqCrrCΦpCrrΦqCrr* (A1)
T4: *CNΦNCpqCNCpqNCpqCΦpCppΦqCqq* (T3, T2)

* T5: $CNPNCpqCPpPq$ (T4, Df. P)
T6: $C\Phi N\Psi NpCqqCqqp$ (A2)
T7: $C\Phi N\Psi NpCNpNpCqqp$ (T6, T2)
T8: $C\Phi N\Psi NpCNpNpCN\Psi NpCNpNpN\Psi NpCNpNpp$ (T8, T2)
* T9: $CPNFNpp$ (T8, Df. P, Df. F)
* RH: $\vdash\alpha \rightarrow \vdash N\Phi N\alpha CN\alpha N\alpha$ (R1)
$\rightarrow \vdash NPN\alpha$ (Df. P).

In proving the U-counterparts of the postulates, we may begin by observing that, since

$$Ta\Phi\alpha\beta = \Sigma bKUbaKTb\alpha\Pi cCUbcCUcaTc\beta$$

and $TbN\alpha = NTb\alpha$,

$$Ta\Phi N\alpha\beta = \Sigma bKUbaKNTb\alpha\Pi cCUbcCUcaTc\beta$$

and

$$\begin{aligned} TaN\Phi N\alpha\beta &= NTa\Phi N\alpha\beta \\ &= N\Sigma bKUbaKNTb\alpha\Pi cCUbcCUcaTc\beta \\ &= \Pi bCUbaNKNTb\alpha\Pi cCUbcCUcaTc\beta \\ &= \Pi bCUbaCNTb\alpha N\Pi cCUbcCUcaTc\beta \\ &= \Pi bCUbaC\Pi cCUbcCUcaTc\beta Tb\alpha. \end{aligned}$$

Hence, for the U-counterpart of R1, we have

$\vdash Ta\alpha \rightarrow \vdash Tb\alpha$ (substitution; α, being a purely tense-logical formula with no a's in it, will be unaltered)

$\rightarrow \vdash CUbaC\Pi cCUbcCUcaTc\beta Tb\alpha$ (by $CpCqCrp$)
$\rightarrow \vdash \Pi bCUbaC\Pi cCUbcCUcaTc\beta Tb\alpha$ (by U.G.)
$= \vdash TaN\Phi N\alpha\beta$.

The U-counterpart of R2 is

$$\vdash TaC\alpha\beta \rightarrow \vdash TaC\Phi\gamma\alpha\Phi\gamma\beta.$$

Of this, the antecedent $= \vdash CTa\alpha Ta\beta = \vdash \Pi fCTf\alpha Tf\beta$ (by U.G.), which last we shall import as an antecedent in proving the consequent of the rule. This consequent may be expanded as follows:

$$\begin{aligned} &TaC\Phi\gamma\alpha\Phi\gamma\beta \\ &\quad = CTa\Phi\gamma\alpha Ta\Phi\alpha\beta \\ &\quad = C\Sigma bKUbaKTb\gamma\Pi cCUbcCUcaTc\alpha \\ &\qquad -\Sigma dKUdaKTd\gamma\Pi eCUdeCUeaTe\beta \\ &\quad = \Pi bCUbaCTb\gamma C\Pi cCUbcCUcaTc\alpha \\ &\qquad -\Sigma dKUdaKTd\gamma\Pi eCUdeCUeaTe\beta \end{aligned}$$

This last we prove as follows:

ΠbC	(1)	$\Pi fCTf\alpha Tf\beta$ (imported from antecedent of rule)	
C	(2)	Uba	
C	(3)	$Tb\gamma$	
C	(4)	$\Pi cCUbcCUcaTc\alpha$	
K	(5)	$\Pi eCUbeCUeaTe\alpha$	(4)
K	(6)	$\Pi eCTe\alpha Te\beta$	(1)
K	(7)	$\Pi eCUbeCUeaTe\beta$	(5, 6)
K	(8)	$KUbaKTb\gamma\Pi eCUbeCUeaTe\beta$	(2, 3, 7)
	(9)	$\Sigma dKUdaKTd\gamma\Pi eCUbeCUeaTe\beta$	(8, E.G.).

Here the steps from (4) to (5) and from (1) to (6) are made by the re-lettering of bound variables which is directly or consequentially permitted in all normal systems of quantification theory ($E\Pi x\phi x\Pi y\phi y$). There will be similar steps in the proof of A1. For this axiom we have, to begin with,

$$
\begin{aligned}
Ta\ (\mathrm{A1}) &= TaCN\Phi NCpqrC\Phi pr\Phi qr \\
&= CTaN\Phi NCpqrCTa\Phi prTa\Phi qr \\
&= CTa\Phi prCTaN\Phi NCpqrTa\Phi qr \\
&= C\Sigma bKUbaKTbp\Pi cCUbcCUcaTcr \\
&\qquad -CTaN\Phi NCpqrTa\Phi qr \\
&= \Pi bCKUbaKTbp\Pi cCUbcCUcaTcr \\
&\qquad -CTaN\Phi NCpqrTa\Phi qr.
\end{aligned}
$$

And this, using in the second line the expansion of $TaN\Phi N$ worked out above,

$$
\begin{aligned}
&= \Pi bCKUbaKTbp\Pi cCUbcCUcaTcr \\
&\qquad -C\Pi dCUdaC\Pi eCUdeCUeaTerCTdpTdq \\
&\qquad -\Sigma fKUfaKTfq\Pi gCUfgCUgaTgr,
\end{aligned}
$$

which we prove as follows:

ΠbC	(1)	Uba	
C	(2)	Tbp	
C	(3)	$\Pi cCUbcCUcaTcr$	
C	(4)	$\Pi dCUdaC\Pi eCUdeCUeaTerCTdpTdq$	
K	(5)	$CUbaC\Pi eCUbeCUeaTerCTbpTbq$	(4, U.I.)
K	(6)	$\Pi eCUbeCUeaTer$	(3)
K	(7)	Tbq	(5, 1, 6, 2)
K	(8)	$\Pi gCUbgCUgaTgr$	(3)
K	(9)	$KUbaKTbq\Pi gCUbgCUgaTgr$	(1, 7, 8)
	(10)	$\Sigma fKUfaKTfq\Pi gCUfgCUgaTgr$	(9, E.G.).

Finally, for A2, $C\Phi N\Psi Npqqp$, note first that the antecedent

$$Ta\Phi N\Psi Npqq$$
$$= \Sigma bKUbaKTbN\Psi Npq\Pi cCUbcCUcaTcq.$$

And within this (using the analogue of the form we found above for $TaN\Phi N$),

$$TbN\Psi Npq$$
$$= \Pi dCUbdC\Pi eCUbeCUedTeqTdp.$$

From the antecedent expanded at this point also, we have to prove Tap, and do so as follows:

ΠbC	(1)	Uba	
C	(2)	$\Pi dCUbdC\Pi eCUbeCUedTeqTdp$	
C	(3)	$\Pi cCUbcCUcaTcq$	
K	(4)	$CUbaC\Pi eCUbeCUeaTeqTap$	(2, U.I.)
K	(5)	$\Pi eCUbeCUeaTeq$	(3)
	(6)	Tap	(4, 1, 5).

7. *Geach's definitions of Kamp's constants.* It should be added to all this, however, that if we are prepared to use the techniques and assumptions which enable Geach to define F in terms of the Diodorean M (see last chapter), we *can* define Φ and Ψ in terms of P and F. That is, we can do it if we allow ourselves (*a*) the use of propositional quantifiers, and (*b*) some such assumption as that at each instant there is something which is true at that instant only. Kamp's Φpq, 'It was the case that p, and it has been the case that q from then till now', clearly implies Miss Anscombe's function Tpq, 'It was the case that p and then q', definable as $PKPpq$. It is not, on the other hand, implied by this ('p and then q' does not imply 'p and then q-ever-since'). On the other hand, Φpq is implied by, but does not imply, the conjunction of Tpq and $NTTpqNq$ ('We have had p-and-then-q, but have *never* had p-and-then-q-and-then-not-q'). But if at each time when p is true there is a proposition (say r) which is true at that time only, then 'p and q-ever-since' does imply (as well as being implied by) 'For some r, it has been that p-and-r and then q, and it has never been that p-and-r and then q and then not q', i.e.

$$\Sigma rKTKprqNTTKprqNq.$$

This definition adapts an analogous direct definition of '*p* the last time that *q*' given by Geach (1966). In view of his previously mentioned definition of *F* in terms of the Diodorean *M*, this means that the whole of tense-logic in Φ and Ψ, and not only the *P–F* fragment of this, may be developed in terms of Diodorean possibility and its past-tense counterpart, with propositional quantifiers. What *postulates* we require for such development over and above the known postulates for the Diodorean *M* and its image (i.e. 'Hamblin's original system', of Chapter IV), and ordinary postulates for quantification, is not fully known; nor is it clear what is the *weakest* system that can be obtained in this way.

The last question is important, because when it is said that in using Geach-style definitions we must 'assume' that, say, time is non-circular, it is not meant that Geach-style definitions will only give us, say, K_t, if it is laid down axiomatically that time is non-circular. The position is rather that if we use such definitions we shall be able easily to prove equivalences (e.g. $EFp\Sigma qKNqMKpq$, $E\Phi pq\Sigma rKTKprqNTTKprqNq$) which are only plausible if the non-circularity assumption is made; i.e. we cannot build in this way systems which are so weak as to be non-committal on this point.

One final piece of pure speculation: in constructing a logic of measured intervals within a 'Φ–Ψ logic', supplemented by suitable and plausible assumptions about origin-events and periodic processes, it may well be necessary to consider the relevance of relativistic physics, and this may result in a rather different type of *Pn–Fn* logic from that sketched earlier in this chapter. But this is a development which I am not at all competent to pursue, and the remaining two chapters will be concerned with complications connected with philosophical problems of a more traditional kind.

Note. If we define $\Omega 1qp$, '*p* the last time that *q*', as $\Phi KqpNq$, we may define Ωnqp, '*p* the *n*th time ago that *q*', as $\Omega 1q\Omega(n\text{-}1)qp$, '(*p* the *n*-1th time ago that *q*) the last time that *q*'.

VII

TIME AND DETERMINISM

1. *Arguments for the incompatibility of foreknowledge (and fore-truth) and indeterminism*). THE eighteenth-century American philosopher Jonathan Edwards, in his *Enquiry* into the freedom of the will, has a simple argument to show that God's foreknowledge is just as inconsistent with a real contingency in future events as his directly foreordaining them would be.[1] In an earlier part of this work,[2] he had observed that there are three ways in which 'the subject and predicate of a proposition' may have such a 'full, fixed, and certain connexion' as to make the 'thing affirmed' in that proposition 'necessary'. He mentions first something like logical necessity: 'it may imply a contradiction to suppose them not connected.' Then—and this is going to be important—'the connexion of the subject and predicate of a proposition, which affirms the existence of something, may be fixed and made certain, because the existence of that thing is already come to pass; and either now is, or has been; and so has as it were made sure of existence. . . . Thus the existence of what is already come to pass is now become necessary; '*tis become impossible it should be otherwise than true, that such a thing has been*' (italics mine). Thirdly, there may be a *consequential* necessary connexion between the subject and predicate. 'Things which are perfectly connected with other things that are necessary, are necessary themselves, by a necessity of consequence.' Edwards notices at this point that 'all things which are future, or which will hereafter begin to be, which can be said to be necessary, are necessary only in this last way'. If their existence were 'necessary in itself', they 'always would have existed', and *ex hypothesi* they have *not* 'already come to pass'. So 'the only way that anything that is to come to pass hereafter, is or can

[1] Jonathan Edwards, *A Careful and Strict Enquiry into the Modern Prevailing Notions of that Freedom of Will which is supposed to be Essential to Moral Agency, Virtue and Vice, Reward and Punishment, Praise and Blame* (1764), Part II, Section xii, subsection i.

[2] Part I, Section iii.

be necessary, is by a connexion with something that is necessary in its own nature, or something that already is, or has been: so that the one being supposed, the other certainly follows'. He might have added, surely, that what is a necessary consequence of something 'necessary in itself' would also 'always have existed', so that it is only by necessary connexion with what 'has already come to pass' that what is still merely future *can* be necessary.

That way, however, it also *must* be necessary, and this is the nerve of his later argument about foreknowledge. 'I observed before', he says, 'that in things which are past, their past existence is now necessary . . . 'tis too late for any possibility of alteration in that respect: 'tis now impossible, that it should be otherwise than true.' That's his Point (1). Point (2) is that if there is such a thing as a 'divine foreknowledge of the volitions of free agents' (the paradigm case, in all these discussions, of supposedly contingent future events), then 'that foreknowledge . . . is a thing which already *has*, and long ago *had* existence; and so . . . it is now utterly impossible to be otherwise, than that this foreknowledge should be, or should have been'. Point (3): 'Those things which are indissolubly connected with other things that are necessary, are themselves necessary. As that proposition whose truth is indissolubly connected with another proposition, which is necessarily true, is itself necessarily true'. This is the modal formula *CLCpqCLpLq*. And Point (4): 'if there be a full, certain and infallible foreknowledge of the future existence of the volitions of moral agents, then there is a certain and indissoluble connexion between those events and that foreknowledge.' Being known necessarily implies being true. Therefore, 'by the preceding observations, those events are necessary events; being infallibly and indissolubly connected with that whose existence already is, and so is now necessary, and can't but have been'.

Edwards insists that in his Part (4) he is *not* saying that God's foreknowledge *causes* things to happen, any more than his 'after-knowledge' does. 'Infallible Foreknowledge may *prove* the Necessity of the event foreknown, and yet not be the thing which *causes* the Necessity.' Edwards further argues, I think with some cogency as well as ingenuity, that if 'God's Foreknowledge is not the cause, but the effect of the existence of the event foreknown, this is so far from shewing that this Foreknowledge does not infer' (i.e. prove) 'that Necessity of the existence of that event, that it rather shews the contrary the more plainly. Because it shews the existence of the event to be so settled and firm, that it is as if it had already been; . . . its future existence has already had actual *influence* and *efficiency*, and has *produced an effect*, Prescience: the effect exists already, and as the effect supposes the cause, . . . and depends entirely upon it, therefore it is as if the future event, which is the cause, had existed already.'

The logical terminology of these passages is a little antiquated; there's too much about subjects and predicates in it, and too much talk of events as 'existing' rather than happening. But the broad pattern of it is powerful. Nor was Edwards the first to invent it. In discussing 'Whether God knows singular future contingents', Aquinas[1] mentions an objection to the proposition that he does, which runs as follows: Given any true proposition of the form 'If p then q', if the antecedent p is absolutely necessary, the consequent q is absolutely necessary. The phrase *est necessarium absolute* does not here mean quite the same as Edwards's first, more or less logical, kind of necessity. It means that q does not just appear as a component of a necessary implication, but is itself a necessary truth (in whatever sense of 'necessary' may be relevant). The schoolmen made a distinction here between *necessitas consequentiae*, necessity of the implication, and *necessitas consequentis*, necessity of the implied proposition. The form 'If p then necessarily q' need not mean that from the truth of p it follows that q is itself a necessary truth, i.e. it need not mean 'If p then necessarily-q'; it may only mean that from the truth of p (which could quite well be the *contingent* truth of p) the truth of q (which again could quite well be the contingent truth of q) *necessarily follows*, i.e. it may mean 'If p then-necessarily q'. This is not really a necessity of q at all, but only a necessary connexion between q and something else. Proponents of the argument we are now considering are sometimes charged with confusing these two senses of 'If p then necessarily q'; but the charge is groundless. They have in fact usually been perfectly well aware of the distinction; what they are exploiting is a certain logical relation that does exist between the two sorts

[1] *De Veritate*, Q. 2, Art. 12, Obj. 7. For a fuller analysis of this argument see A. Prior, 'The Formalities of Omniscience', *Philosophy*, April 1962.

of necessity (or the two points at which the necessity may operate), namely that where *not only the implication as a whole, but also the implying proposition, is necessarily true*, there the implied proposition is necessarily true also. This is of course the modal thesis *CLCpqCLpLq* again. Where authorities are cited here, the appeal is generally to Aristotle's *Anal. Pr.* 34^a23 or *Anal. Post.* 75^a4–12.

The objector's second premiss is that 'If anything is (already) known to God' (*est scitum a deo*), 'then that thing will be.' But the antecedent of this, at least if it is true at all, is necessary, if only because God already *has* come to know the thing, so that nothing can now make him not have known it—*quod fuit, non potest non fuisse* (what has been, cannot now not have been). And so—the corollary is too obvious for Thomas to bother drawing it explicitly—what God already knows will happen *isn't* now contingent at all. Where authority is cited here, the appeal is usually to Aristotle's *Nicomachean Ethics* 1139^b8 ff. and to his *De Caelo* 283^b12.

It is the foreknowledge as such that is incompatible with contingency by this argument; that it is God's foreknowledge, is immaterial. In Cicero's *De Fato*,[1] the same point is made in connexion with astrological principles such as 'If anyone is born under the Dog Star, he will not die at sea'. From this it follows that if Fabius (who is now living) was born under the Dog Star, *he* will not die at sea. But here the antecedent is necessary, since 'all true past-tense propositions are necessary', and so the consequent must be true. This is put forward by Cicero as a kind of argument which Diodorus would use. It does have something of the flavour of the Master-argument; like the latter, it is directed against those who argue that we have no control over the past but think we have some over the future; and in both cases the trick appears to be that of conveying the admitted necessity from the past to the future by means of some proposition that necessarily connects the two.

An astrologer's prophecy is rather a weak support for such a connexion, and indeed Cicero is not here defending the fatalistic conclusion but using it to denounce astrology. And even God's foreknowledge is not as widely accepted nowadays as it was either in Aquinas's Europe or in Edwards's America.

[1] Capp. vi, vii.

But there are principles of tense-logic that can be, and have been, put to the same purpose. The most lucid statement of the tense-logical argument that I know of is that of the fifteenth-century Louvain philosopher Peter de Rivo.[1] His central point is that if, before a given event occurs, statements asserting its future occurrence were already true, we could use this to set up an argument exactly like the ones discussed by Cicero and Aquinas (both of whom de Rivo quotes). For, from the truth, already, of '*X* will be *Y*' it necessarily follows that *X* will be *Y* (there is an appeal here to the 'Tarskian' principle enunciated in Aristotle's *Categories* 14^b13–17); if that were true already its truth would be now beyond prevention (*inimpedibile*), for we have no power over the past (*ad preteritum non est potentia*); but only the unpreventable follows from the unpreventable; so if '*X* will be *Y*' is true already, that *X* will be *Y* is already inevitable.[2] The presentation is a little metalinguistic; but not at all points; e.g. he indicates at one point that what he is attacking (in order to avoid Wyclif's 'execrable determinism') is the view that 'of whatever is now the case it was earlier true that it was going to be the case', *CpPFp*.

There are portions of Aristotle's famous 'sea-battle' chapter (*De Interpretatione*, Ch. 9) which read as if the same argument is being put forward. Certainly Cicero credits Epicurus with the view that in order to escape determinism we must deny that predictions about issues which are still genuinely open are either true or false (de Rivo's conclusion).

2. *Formalization of these arguments.* In trying to formalize these arguments, let us use *L* for an undefined 'necessarily', i.e. *not* for 'is or will be' or for 'is or has been or will be' but for something more like 'now-unpreventably' ('necessary' propositions are those outside our power to make true or false). Then one of the main premisses of these arguments would appear to be

1. *CPpLPp*, 'Whatever has been the case now-unpreventably has been the case'.

[1] The papers in the de Rivo controversy have been collected together, with an excellent introduction, in L. Baudry's *La Querelle des Futurs Contingents* (Louvain 1465–75): *Textes Inédits* (Paris, 1950).

[2] Baudry, op. cit. pp. 70 ff., 80–81, 85–86.

We then proceed thus:

2. $CPFpLPFp$ (1 p/Fp), i.e. 'If it has been that it will be that p, it now-unpreventably has been that it will be that p'.
3. $CFpPFp$, 'Of what will be, it has been the case that it will be'.
4. $CFpLPFp$, 'Of what will be, it now-unpreventably has been the case that it will be'. (2, 3, syll.)
5. $CLCpqCLpLq$.

And now, if we have something like

6. $LCPFpFp$, 'Necessarily if it has been the case that it will be, it will be',

we could go by 5 and 6 to

7. $CLPFpLFp$,

and from this and 4 to the fatalistic conclusion

8. $CFpLFp$.

But this formalization won't do, as 6 is plainly false, and so would be its counterpart in the theological version of the argument, 'Necessarily if it has been the case that God knows that it will be the case, it will be the case', or more colloquially, 'If God knew that it would be the case, it will be the case'. This is false, i.e. as a law, simply because by the time of utterance what was going to happen, or what God knew would happen, may have already happened, and it may not be going to happen again. Cicero in using his example about the Dog Star was sufficiently aware of this problem to suppose the argument to be going on before Fabius had already died and thereby already fulfilled or falsified the prophecy.

What we really want to say, at the point where 6 has been put, is that its having been the case some time before now that it would be the case a *longer* time later (e.g. its having been the case yesterday that I was going to smoke two days later) necessarily implies that it will now be the case not quite so much later (in the example, that I will be smoking tomorrow). Something from metric tense-logic would give us what we want here, namely $LCPmF(m+n)pFnp$. Given this, with corresponding modification of the other formulae, we have

1. *CPmpLPmp*
2. *CPmF(m+n)pLPmF(m+n)p* (1, subst.)
3. *CFnpPmF(m+n)p*
4. *CFnpLPmF(m+n)p* (3, 2, syll.)
5. *CLCpqCLpLq*
6. *LCPmF(m+n)pFnp*
7. *CLPmF(m+n)pLFnp* (5, 6)
8. *CFnpLFnp* (4, 7, syll.)

In this version, the tense-logic is less questionable (though it *can* be questioned, as we shall see), but it might be said that the relation affirmed in 1 between necessity and the past is not quite the one being asserted by the propounders of the argument. What the propounders of the argument are ascribing to the past, it may be said, is a kind of necessity which is or entails unalterableness. Things may indeed become 'necessary' in this sense which were not so before; decisions, or the mere march of events, may close possibilities which were formerly open; we may say that a thing is now necessary because it is 'too late' for it to be otherwise—it has as it were 'lost its chance' of being false—but once this happens, it has happened for good and all; to say that a thing's being thus and so has become necessary is to say that from now on it must stay that way. But the past is only unchangeable in the sense that what has been the case will-always have been the case. It is not unchangeable, as we have already seen, in the sense that once a certain proposition, say that 'there will be a sea-battle a day hence', has come to be true, that proposition is bound to stay true. If that proposition was true yesterday, what is bound to be true today is not that there will be a sea-battle a day hence but that there *is* a sea-battle *today*.[1] Nor is the past unchangeable in the sense that if something was the case the interval *n* ago (say this time yesterday), then it will always be the case that it was the case the interval *n* before. ('I had sausages for breakfast yesterday' may be true today and false tomorrow.) Even if we do have *CPpGPp* (and so *CPFpGPFp*), we not only don't have *CPFpGFp*, but don't even have *CPnpGPnp*. (This is McTaggart's objection to the dictum that the past does not change.) But what our new law 1 states is that if it was the case the interval *n* ago

[1] This point was made by Suarez.

that p, then it is now necessary that it was the case the interval n ago that p. It is 'necessary'—and yet the least little bit later it is perhaps not even true, the truth then being not Pnp but $P(m+n)p$ (where m is that least little bit). If there is any necessity here, it is, or at least is liable to be, a quite momentary one, and what sort of necessity would this be?

I am inclined to think, however, that this objection is frivolous. The change in truth-value that is mentioned here is itself inevitable; it is not something that we by our choice, or some chance turn of events, can bring about; and it does not alter the fact that at each instant what happened the interval n before cannot *then* not have happened the interval n before. There may, all the same, be other objections to the law 1, in both its forms.

Perhaps the argument comes through most intuitively of all in a mixed tensed and dated calculus in the style of Rescher. Suppose we again use the form Tap for 'It is true at date a that p', with the postulates

$$\text{RT: } \vdash\alpha \rightarrow \vdash T a\alpha;$$

and

$$\text{TC: } CTaCpqCTapTaq,$$

from which we can derive the rule

$$\text{RTC: } \vdash C\alpha\beta \rightarrow \vdash CTa\alpha Ta\beta.$$

We add to this (following Rescher) the form Dap for 'It is *determined* at a that p, $DaFnp$ expressing the *pre*-determination of p and $DaPnp$ its post-determination. For D we have the following laws:

$$\text{RD: } \vdash\alpha \rightarrow \vdash Da\alpha$$

$$\text{DC: } CDaCpqCDapDaq,$$

from which we get

$$\text{RDC: } \vdash C\alpha\beta \rightarrow \vdash CDa\alpha Da\beta,$$

and we also have

$$\text{DP: } CTaPnpDaPnp.$$

This (if it is true at a that it was the case n ago that p, it is determined at a that it was the case, etc.) is the usual law of universal post-determination (*quod fuit, non potest non fuisse*). From this we can prove universal *pre*-determination as follows:

1. $CFnpPmF(m+n)p$ (from tense-logic)
2. $CTaFnpTaPmF(m+n)p$ (1, RTC)
3. $CTaFnpDaPmF(m+n)p$ (2, DP, syll.)
4. $CPmF(m+n)pFnp$ (from tense-logic)
5. $CDaPmF(m+n)pDaFnp$ (4, RDC)
6. $CTaFnpDaFnp$ (3, 5, syll.).

3. *The classical answers to these arguments.* In ancient accounts of the Diodorean Master-argument and of the reception it met with, we are told that one Stoic logician, Cleanthes, was driven by it to deny that past-tense truths are always necessary, while another, Chrysippus, was driven to deny that the impossible cannot follow from the possible. In reacting to the argument we are now considering, some have followed Cleanthes, and others have denied the tense-logical principle that if ever '*S* is *P*' *is* true, then '*S* will be *P*' formerly *was* true. The first line was taken notably, in the Middle Ages, by William of Ockham, who said that the principle that what has been cannot now not have been only applies to past-tense propositions which are not equivalent to future-tense ones (in the way in which 'It was the case yesterday that it would be the case two days later that I-am-smoking' is equivalent to 'It will be the case tomorrow that I am smoking').[1] The fifteenth-century critics of Peter de Rivo, notably Ferdinand of Cordova, put a similar proviso on the principle *ad preteritum non est potentia*, and argued that we *do* have some power over that much of the past which consists in the past truth of future-tense propositions.[2] (By deciding whether to smoke or not to smoke tomorrow, I decide whether or not to make it have been true yesterday that I would smoke two days later.)

The other line, that a thing's being the case today does *not* imply that it *was* true yesterday that it would be the case a day later, was taken by Aquinas and de Rivo; and among the ancients, according to Cicero, it was taken by Epicurus; and according to many, it was taken before that by Aristotle. The ancient and medieval proponents of the second alternative did not say that before a future event was 'already present in its causes' (as Aquinas put it), it would have been *false* to say

[1] Ockham, *Tractatus de Praedestinatione* (Franciscan Institute edition, 1945), p. 6.
[2] Baudry, op. cit. p. 159.

that it was going to happen, but rather that it would have been *neither true nor false* to say this. 'With what is now the case', Peter de Rivo says, 'it need not have been previously either true or false to say that it was going to be the case.'

4. *Formalization of the Ockhamist answer.* I propose now to take each of these escape-routes in turn (though I shall modify the second a little), and see how it can be formalized; and will begin with the first solution, i.e. the Ockhamist one. In saying that the *rule* that truths about the past are necessary only applies to those past-tense propositions which are not equivalent to future-tense ones, Ockham is not saying that past-tense propositions which are equivalent to future-tense ones are *never* necessary. They would presumably be necessary at least if the equivalent future-tense ones were, e.g. if *FnCpp* is a necessary truth, so presumably is *PmF*$(n+m)$*Cpp*. But only past-tense propositions which are not logically equivalent to future-tense ones are, so to speak, necessary in virtue of their pastness. It is a little difficult, however, to put this into a law. We are trying to lay down postulates of which the purpose is precisely to help us find out what is logically equivalent to what; Ockham's rule only seems to be operable when this is already done; but it is one of the things we need to be able to use in finding out what *are* the laws of the system.

Still, there *is* something about the very structure of a past-tense proposition equivalent to a future-tense one which does enable us to see whether a given past-tense proposition could fall into this category or not. Curious special cases apart (e.g. the plain *PnCpp* being equivalent to *FnCpp* because both express logical laws), past-tense propositions are equivalent to future-tense ones only if they have a subordinate future-tense clause within them, as in ⊢*EPmF*$(m+n)$*pFnp*. Even so, it is not easy to lay down a law for past-tense propositions which will exclude even these ones. The plain ⊢*CPpLPp* or ⊢*CPnpLPnp*, for example, does not itself have any future-tense operators in it, but cannot express the restricted law we want, since we can immediately *put* future-tense operators in it by substitution for *p*; indeed, with free substitution of propositional formulae for propositional variables, how can we possibly keep them out?

Restrictions on substitution-rules, however, are not impossible to operate, and there is a further very strong reason for believing that they are hardly avoidable here. Ancient and medieval writers who have laid it down that we have no power over the past (and Edwards later on too), have generally made a similar remark about the present; we may cite here Aristotle's much-discussed remark in his 'sea-battle' chapter that 'Whatever is, when it is, is necessarily, and whatever is not, when it is not, necessarily is not.' It is the future only which is 'open both ways'. But if we just lay down ⊢*CpLp* without any restrictions on substitution, we will have a much quicker proof that the future is necessary too, than any we have yet given; it will just be *CpLp* with the substitution *p*/*Fp*. Under these conditions, the necessity-operator *L* in fact becomes quite vacuous.

Accepting, then, that we must restrict substitution-rules, the restriction must be based on the division of propositions into two classes—on the one hand, those which as it were challenge comparison with the world as it already is, and which we cannot possibly make true or false by any decision that is now open to us, because they express the given situation in which any decisions of ours must be made; and on the other hand there are propositions, like 'Eclipse will win', which so look beyond the present to the future that they must as it were lie on the table until the race is run. This is not to mean, with respect to the latter class of propositions, that they are not yet either true or false; but their 'wait and see' character so infects whatever compounds they enter into that the present-tense assertion that such a proposition is now true has itself this 'wait and see' character and must just lie on the table until the verifying event occurs; and ditto statements before now that the thing would happen after now.[1]

One simple way of restricting substitution is to use one sort of proposition variables, say the usual *p*, *q*, *r*, etc., to stand for propositions of all the kinds that the system contains (in this case, both for those which we cannot now make true or false and for the 'wait and see' ones which we sometimes can), and another sort of propositional variables, say *a*, *b*, *c*, etc., only for

[1] At this point, and quite generally in my understanding of the position that I have called 'Ockhamist', I am very much indebted to discussions with J. M. Shorter in 1957–8. Shorter has convinced me, in particular, that the very non-standard semantics which are said on pp. 94–95 of *Time and Modality* to be involved in the Ockhamist position of G. Ryle, are not so involved.

propositions with a particular internal structure.[1] Here we shall use the restricted variables only for propositions expressing what Peter de Rivo calls the 'now-unpreventable', i.e. propositions which in general have no trace of futurity in them. We use the term 'formulae' to cover all the propositional formulae of the system, and define these inductively as follows:

(1) Propositional variables (of both sorts) are formulae.
(2) If α and β are both formulae, so are $\mathcal{N}\alpha$, $C\alpha\beta$ ($K\alpha\beta$, etc.), $\Pi n\alpha$, $\Sigma n\alpha$, $Pn\alpha$, $Fn\alpha$, and $L\alpha$.
(3) There are no others.

We use the term 'A-formulae' to cover only a subclass of these, defined as follows:

(1) A-variables (i.e. *a*, *b*, *c*, etc.) are formulae.
(2) If α and β are both A-formulae, so are $\mathcal{N}\alpha$, $C\alpha\beta$ ($K\alpha\beta$, etc.), $\Pi n\alpha$, $\Sigma n\alpha$, and $Pn\alpha$.
(3) If α is any formula, $L\alpha$ is an A-formula.
(4) There are no others.

Pna, for example, is an A-formula by clause (2) of the definition, but *Fna* is not; nor, consequently, is *PmFna*. On the other hand, *LFna* and even *LFnp* are A-formulae; that something (even something future) is now-unpreventable, is itself (when true) now-unpreventable.

Even with this last bit of liberality, the conditions on the formation of A-formulae might be thought to be too restrictive. For example, *P*(*n*+*m*)*Fma* is not an A-formula by our definition, although it is not equivalent to any future-tense formula but rather to the simple past-tense *Pna*, so that we *do* want, in an Ockhamist logic, to have *CP*(*n*+*m*)*FmaLP*(*n*+*m*)*Fma*. But although we cannot directly obtain this by substitution in the law *CaLa* for A-formulae, we shall find that it is easily derivable in the system in other ways, and similarly with other formulae which are not themselves A-formulae but are logically equivalent to these.

We lay it down, then, that any formula may be substituted throughout a thesis for one of the unrestricted propositional variables *p*, *q*, *r*, etc., and that only A-formulae may be sub-

[1] For an earlier use of this technique, applied to a different problem, see *Time and Modality*, App. B. Cf. also here, ch. V, Section 6.

stituted for the A-variables *a*, *b*, *c*, etc. We then take over the whole of the metric tense-logic of the third sort discussed in the last chapter, i.e. with *Pn* and *Fn* both primitive and with interval-measures restricted to positive numbers, and the whole of it still formulated with unrestricted variables; *except* that the mirror-image rule is restricted to formulae not containing *L*; and we add the following postulates for *L* ('now-unpreventable'):

RL: $\vdash\alpha \rightarrow \vdash L\alpha$

L1. *CLpp* — LF: *CLFnpFnLp*

L2. *CLCpqCLpLq* — LFΠ: *CΠnFmLFnpFmΠnLFnp*

L3. *CNLpLNLp* — LPΠ: *CΠnFmLPnpFmΠnLPnp*

LA: *CaLa*

RL, L1, L2, and L3 give us for this undefined *L* the modal system S5. Substitution in LA will give us, e.g. *CPnaLPna*, but not *CFnaLFna* or *CPmFnaLPmFna*. However, if any formula β is logically equivalent to any A-formula α, we can prove $C\beta L\beta$ as follows:

1. $C\alpha\beta$ (hyp.)
2. $C\beta\alpha$ (hyp.)
3. $C\alpha L\alpha$ (LA, subst.)
4. $CL\alpha L\beta$ (1, RL, L2)
5. $C\beta L\beta$ (2, 3, 4, syll.).

In a case of the type just mentioned, we also have $C\alpha L\beta$ (by 3 and 4). For example, let α be the simple A-variable *a* and β be *PnFna*. For our 2 and 1 we then have *CPnFnaa* and *CaPnFna*, which are provable by substitution in the mirror images of FP3 and its converse. Hence we can prove *CaLPnFna*, e.g. if I am now smoking, it now-unpreventably was the case this time yesterday that I would be smoking a day later. On the other hand, there is no way of proving *CaPnLFna*, which would assert (with the same *a*) that if I am now smoking then it was the case this time yesterday that I then-unpreventably would be smoking a day later; and it would of course be intuitively awkward if we *could* prove this.

An alternative formalization would be one in which the only propositional variables are A-variables, formulae and A-formulae being defined as before (except that the first clause in the definition of 'formula' only refers to one type of variable).

The substitution-rule would then be to replace variables by A-formulae, and LA would be the only axiom in the strict sense of a single formula laid down axiomatically; the rest would be replaced by the corresponding axiom *schemata*, e.g. FP1 by the schema *CFnPn*$\alpha\alpha$ and L1 by *CL*$\alpha\alpha$, it being understood with each schema that all results of putting formulae (of any sort) in the place of Greek letters are axioms; e.g. *CLFnaFna* and *CFnPnFnaFna* are axioms. The theses of the modified system would be all the theses of the original one which only use A-variables, including ones obtained in the original system by substitution in formulae using the unrestricted variables.

For the Ockhamist system in this second form, we may define an Ockhamist *model* as a line without beginning or end which may break up into branches as it moves from left to right (i.e. from past to future), though not the other way; so that from any point on it there is only one route to the left (into the past) but possibly a number of alternative routes to the right (into the future). In each such model, formulae are assigned truth values (truth or falsehood) in accordance with the following prescriptions:

(1) Each propositional variable is arbitrarily assigned a single truth-value at each point.
(2) A prima-facie assignment to *Fn*α at a given point x for a given route to the right of x, gives it the value assigned to α at the distance n along that route from x. (If the line branches within this distance, there may be different prima-facie assignments to *Fn*α at x.)
(3) The prima-facie assignment to *Pn*α at a given point x for a given route for α to the right of x, gives it the value assigned to α, for that route, at the distance n to the left of x. From the latter point as far as x, the only rightward route for α which is considered is the one that passes through x.
(4) The assignment to *L*α at x gives it truth if α is given truth in all its prima-facie assignments at x; otherwise falsehood.
(5) Truth-functions and quantifications as usual.

A formula is verified by an Ockhamist model if all actual and prima-facie assignments to it in the model give it truth; and we might define the Ockhamist system as consisting of those formulae which are thus verified by all Ockhamist models. Whether the postulates earlier listed yield all such formulae, i.e. whether they are complete for Ockhamist tense-logic, is not known.

To illustrate the use of an Ockhamist model, consider the following portion of one,

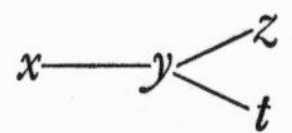

where $xy = m$, $yz = yt = n$, and the proposition a is true at x, y, and z and false at t. Because a is true at z, the prima-facie value of $F(m+n)a$ at x for the route xyz is truth; and that of $PmF(m+n)a$ at y for the route yz for $F(m+n)a$ beyond y, is also truth. But because a is false at t, the prima-facie value of $F(m+n)a$ at x for the route xyt is falsehood, and that of $PmF(m+n)a$ at y for the route yt for $F(m+n)a$ beyond y is also falsehood. Hence the assignment to $LF(m+n)a$ at x, and that to $LPmF(m+n)a$ at y, are both falsehood. $CFnpLPmF(m+n)a$ is therefore false at y on the assignment for the route xyz; since Fnp is true at y using this route, while $LPmF(m+n)a$ is simply false.

On the other hand, since Fna is assigned truth at y for the route yz, $PnFna$ is true at z regardless of what happens to y beyond z, for at the distance n to the left of z, i.e. at y, Fna is assigned truth for the only route from y which passes through z. The only value assigned to $PnFna$ at z is therefore truth, so that we can assign truth at this point to $LPnFna$ also, and to $CaLPnFna$. On the other hand, $LFna$ is false at y, (since Fna has one prima-facie assignment of falsehood there, namely that using the route yt), and $PnLFna$ therefore false at z, and $CaPnLFna$ false there too.

5. *Ultimately converging time.* Before passing on to the alternative to the Ockhamist system, it is worth observing that the device of restricting substitution by the use of special variables may be extended to deal with another point. I suggested in an earlier chapter that 'It will all be the same in a hundred years' time, no matter what we do now' cannot be *quite* true, since what we do now will at least make a difference to what *will have been the case* by then. But people who make this sort of remark may well complain that that's not the sort of thing they intend it to

apply to. Indeed, what is normally meant would exclude a great deal more than this. But even if it is intended quite sweepingly—even if what is meant is that whatever free play we may now have, all that is to happen after a certain time is *quite* fixed—it *must* be understood as not applying to the future truth of past-tense propositions, or there just couldn't *be* any 'free play' in the meantime.

The logical problem involved here is exactly analogous to that involved in developing a precise concept of post-determination which will not entail predetermination. It will not quite do to say that it is already determined what present-tense and future-tense propositions will be true after a certain time, though we have some choice as to what past-tense propositions will be true then; for the future-tense propositions of that time will include ones like 'It will be the case tomorrow that it was the case 50 years ago that *p*', and maybe we don't want to say that it is now quite determined which of *those* will be true; while on the other hand the past-tense propositions of that time will include ones like 'It was the case 50 years ago that it would be the case 500 years later that *p*', and we *do* want to say that it is now determined which of *those* are to be true; and as to the present-tense propositions of that time, these could include *all* of them, since 'It *is* the case that—' is prefixable to *anything*. We need to formulate the thesis of remote predetermination in terms of a class of propositions which are 'non-past' in much the same way as our A-formulae above are 'non-future'.

6. *Formalization of the Peircean answer, and comparison with the Ockhamist.* Turning now to the other way of answering the argument from post-determination to predetermination, that of denying that Fnp always implies $PmF(n+m)p$, I begin by modifying the ancient and medieval presentation of this alternative at one point. What is said by writers like Peter de Rivo is that predictions about an as yet undetermined future are neither true nor false. It did seem to me in the early 1950s that this was the only way to present an indeterminist tense-logic, but in *Time and Modality* two alternatives to this were mentioned, one the Ockhamist position developed in Ryle's *Dilemmas* (which, however, I misrepresented) and the other the alternative which I now want to pursue further. What here takes the

place of a third truth-value is a sharp distinction between two senses of 'It will not be the case the interval *n* hence that *p*'. This may mean either

(A) 'It will be the case the interval *n* hence that (it is not the case that *p*)', i.e. *FnNp*;

or

(B) 'It is not the case that (it will be the case the interval *n* hence that *p*)', i.e. *NFnp*.

'Will' here means 'will definitely'; 'It will be that *p*' is not true until it is in some sense *settled* that it will be the case, and 'It will be that not *p*' is not true until it is in some sense settled that not-*p* will be the case. If the matter is not thus settled, both these assertions, i.e. *Fnp* and *FnNp*, are simply false. The weak form (B) can therefore be true for two quite different reasons; it may 'not be the case that *p* will be the case' at the time stated, *NFnp*, because it is already settled beyond any possibility of reversal that it *will* be *not*-the-case; or that it *will* be may 'not be the case' yet simply because it isn't yet settled either way. There is no question now of denying the Law of Excluded Middle *ApNp*; this still holds even in the special case *AFnpNFnp*; and moreover the allied metalogical 'Law of Bivalence', that every proposition (even 'It will be the case the interval *n* hence that *p*', spoken of something as yet undetermined) is either true or false, is not abandoned either (under the circumstances mentioned, 'It will be the case the interval *n* hence that *p*' is simply false, no matter *how* things turn out later on). Nor is it denied that the Law of Excluded Middle *will* be true in every particular case; we have, e.g. *FnApNp* ('It will be the case tomorrow that either there is a sea-battle going on or there isn't'). What *is* denied is that we always have *AFnpFnNp*, i.e. that it always either will be the case that not *p* or will be the case that *p*.

This position clearly entails some radical modifications of the metric tense-logical system set up in the last chapter. For instance, although we can keep FN1, *CFnNpNFnp*, ('If it will be then that not *p*, it won't be then that *p*'), we have to drop FN2, *CNFnpFnNp* ('If it won't be then that *p*, it will be then that not *p*'). This destroys the proofs of the converses of the remaining axioms involving *F*, and we will need to lay down separately (at least in a first axiomatization) those that do still hold. We need also to watch the relations between *F* and *A* and *K*. Since we still have FC, we can prove *CFnKpqKFnpFnq* and *CAFnpFnqFnApq*, but *CKFnpFnqFnKpq* would seem to require separate assertion and *CFnApqAFnpFnq* no longer holds (e.g. as was observed in the last paragraph, we have *FnApNp* but not *AFnpFnNp*). Also, since we have PN2, *CNPnpPnNp*, as well as PN1, *CPnNpNPnp*, but do not have FN2, the mirror-image rule must go, and the mirror images that hold must be separately asserted. Mixtures of *F* and *P* have a particularly complicated logic; we have *CPmpFnP*$(m+n)$*p* and *CpFnPnp* and their converses but not their mirror images, though we do have the converses of their mirror images, i.e. we do have *CPnF*$(m+n)$*pFmp* and *CPnFnpp*.

Shorter pointed out in 1957 that in the system now being considered, which I shall call 'Peircean' for reasons that I shall give below, the rather strong 'will be' is simply the Ockhamist 'necessarily will be', the Ockhamist 'will be' being untranslatable. We can in fact characterize the Peircean system as that fragment of the Ockhamist system in which there are no variables but A-variables, and *F* does not occur except as immediately preceded by an *L*, which last symbol now becomes redundant and so may be dropped. For example, in O (the Ockhamist system) *CaLFnPna* is provable thus:

1. *CaFnPna*
2. *CLaLFnPna* (1, RL, L2).
3. *CaLFnPna* (LA, 2, syll.),

so that *CpFnPnp* holds in P (the Peircean system). But *CaPnLFna* is not provable in O, and so *CpPnFnp* not in P. Again, we have *CLFnNaNLFna* in O, and so FN1 in P; but not *CNLFnaLFnNa* in O, and so not FN2 in P.

To the Ockhamist, Peircean tense-logic is incomplete; it is simply a fragment of his own system—a fragment in which contingently true predictions are, perversely, inexpressible. The Peircean can only say 'It will be that *p*' when *p*'s futurition is necessary; when it is not necessary but will occur all the same, he has to say that 'It will be that *p*' is false; the sense in which it is true eludes him. But to the Peircean, the Ockhamist seems to treat what is still future in a way in which it would only be proper to treat what *has been* future—he views it as it would be proper to view it from the end of time. For the Peircean *can* give a sense in his own language to *past* Ockhamist futures, provided that they are *far enough* past. He can, that is, give a sense to the Ockhamist 'It was to be', *PmFnp*, and even to 'It was contingently to be', *KPmFnpNPnLFnp*, provided that $m > n$ and that there is not too much future in what is represented by *p*. The former is simply, in the *P* language, $P(m-n)p$, and the latter, $KP(m-n)pNPmFnp$. For example, the Ockhamist 'It was the case two hours ago that Eclipse would win an hour later' is in Peircean just 'Eclipse won an hour ago', and 'It was the case two hours ago that Eclipse *would* win an hour later, but not that he *had* to' is in Peircean 'Eclipse won an hour ago, but it was not the case two hours ago that he would win an hour later'.

The Peircean can, I think, even give instruction in the use of Ockhamist tenses, as these are used, e.g. in betting. (We don't refuse to pay up on the grounds that when the man said 'Eclipse will win' what he said was false—or even on the grounds that what he said was neither true nor false—because the matter was still undecided when he said it.) Using 'WAS' and 'WILL' for the Peircean past and future, and 'was' and 'will' for the Ockhamist, the Ockhamist's

> 'Your statement of an hour ago, "Eclipse will win in an hour's time", was true'

goes into Peircean as

> 'It WAS the case an hour ago that you were saying "Eclipse will win", and now he is winning'.

What cannot be said in Peircean is the Ockhamist's 'It *is* to be' (where this does not mean 'It is bound to be'), i.e. his *Fnp*, or his *PmFnp* where $n > m$. But even of this it can be said in a Peircean metalanguage,

> 'If an Ockhamist is now saying "It will be the case an hour hence that Eclipse is winning", then it WILL (now-unpreventably will) be the case an hour hence that either his statement was true or it was false.'

(The dependent 'was' being defined as before.) This is just a case of the Peircean theorem $CpFnAKPnpqKPnpNq$.

Nor does the Peircean logic *need* to be characterized as a fragment of the Ockhamist one. For it could also be characterized as consisting of all those theses which are verified in all Peircean models, a Peircean model being like an Ockhamist model except that the truth-value assignments are as follows:

(1) and (2): Assignments to variables, and prima-facie assignments to $Fn\alpha$, as in the O model.

(3) The actual assignment to $Fn\alpha$ at x gives it truth if all its prima-facie assignments do; otherwise falsehood.

(4) The assignment to $Pn\alpha$ at x gives it the value actually assigned to x at the distance n to the left of x (on the line connected to x).

(5) Truth-functions and quantifications as usual.

It is difficult to define within Peircean logic a 'necessity' for which we can say that all truths about the past, but not all about the future, are necessary. For the F of this logic only enables us to state such truths about the future as *are* necessary. What we can do is to define a sense of 'possibly will' which is distinguishable from the plain 'will', although the analogous sense of 'possibly was' is not distinguishable from the plain 'was'. Mnp, for 'It possibly will be the case the interval n hence', is simply 'It is not the case that it will be the case the interval n hence that not p', $NFnNp$, which is true if either it definitely WILL be the case that p or the matter is still undecided. But $NPnNp$ is true if and only if Pnp is (we have this by PN1 and PN2). This corresponds less closely to ancient and medieval formulations than to C. S. Peirce's description of the past (with, of course, the present) as the region of the 'actual', the area of 'brute fact', and the future as the region of the necessary and the possible.[1] That is why I call this system 'Peircean'.

7. *The Peircean senses of 'will'*. The GH system which we obtain from Peircean tense-logic by writing $G\alpha$ and $H\alpha$ for $\Pi nFn\alpha$ and $\Pi nPn\alpha$ is axiomatizable, in its own terms, as follows. Subjoin to propositional calculus, with substitution and detachment, the rules to infer $\vdash G\alpha$ and $\vdash H\alpha$ from $\vdash \alpha$, and the axioms

[1] *Collected Papers of C. S. Peirce*, 5. 459 and 6. 368.

A1.1. *CGCpqCGpGq* A1.2. *CHCpqCHpHq*
A2.1. *CGpNGNp* A2.2. *CHpNHNp*
A3.1. *CGpGGp* A3.2. *CHpHHp*
A4.1. *CpGNHNp* A4.2. *CpHNGNp*
A5. *CpCHpCGpGHp*

(the question of denseness, etc. being left open). The only non-standard feature here is of course the absence of any mirror image of A5. The disappearance of this from time-systems with a branching future—in which no possible future is singled out as the actual one, and *Gp* means 'It is true throughout all possible futures'—has already been commented on in Chapter III. It might be thought that more would have to go than this; in particular, if A4.2 is abridged to *CpHFp*, 'What is the case has always been going to be the case', this seems to be one of the first things that ought to go (since we don't have, e.g. *CpPnFnp*, on which that would seem to depend). Just for this reason, however, the *NGN* of A4.2 has *not* been abbreviated to *F*; if we do read *F* as simply an abbreviation for *NGN*, all that *Fp* means is that it *could* come to pass that *p*, i.e. *p* is not false-throughout-all-possible-futures, and if *p* is actually occurring it certainly *has* always been the case that *p* is not false-throughout-all-possible-futures (for it to be occurring, there must always have been some possible future which included it). And what A4.2 requires in the underlying calculus P to prove it is not *CpPnFnp* but the weaker *CpNPnFnNp* (if *p* is the case, then it was not the case *n* ago that it definitely would, *n* later, be false'), which *is* in P. (It is equivalent to *CpPnMnp* in the terminology of the last paragraph.)

The *F*-function which means 'It *definitely* will be that', without going so far as 'It definitely will always be that', and for which *CpHFp* is to be rejected in a Peircean-style system, is not definable in terms of *G*. Something like it, however, could be defined in terms of the Peircean *Fn* and introduced into the *GH* calculus independently. The *NGN* function is, in P, an abridgement of *NΠnFnN*; the other *F* could be *ΣnFn*, which in P is not equivalent to the former, but stronger. Certainly *Σn* = *NΠnN*, but this turns *ΣnFn* into *NΠnNFn*, not into *NΠnFnN*, and in the absence of *CNFnpFnNp* we cannot prove *CΠnNFnpΠnFnNp*, and so not the transposed form *CNΠnFnNpNΠnNFnp*.

Even the Peircean *ΣnFn*, however, does not give us *quite* the *F* we want. If *NGN* is too weak, *ΣnFn* is in one way too strong. It tells us that there is some instant, future to now, such that it is now-unpreventably the case that whatever it is will *then* be the case; what is really wanted is the assertion that the thing is bound to happen some time or other (not that there is some time at which it is bound to happen). We want to say, in other words, 'On every route into the future, there is somewhere a point at which *p* is the case', but not 'There is a distance such that *p* is the case that distance along every route', which is what *ΣnFnp* says. We cannot express what we want in P, because we have no machinery for quantifying over routes. We can do it, in a manner, in O. The Peircean *ΣnFnp* is the Ockhamist *ΣnLFnp* (For some *n*, it is bound to happen *n* hence); what we want is rather the Ockhamist *LΣnFnp* ('It is bound to happen some time'). This won't go into Peircean because that language incorporates the Ockhamist *F* only as that is *immediately* preceded by an *L*.

Once again, however, this does not mean that we *have* to describe the language we are after as a fragment of Ockhamist language. We can say that we are after a *GHF* system (*P* can still be defined as *NHN*) consisting of all formulae that are verified by all *GHF* 'models' of a certain type—infinite branching lines again, and truth-values now assigned, in each model, as follows:

(1) Each variable has an arbitrary assignment of truth or falsehood at each point on the line.
(2) *G*α is assigned truth at *x* if α is assigned truth at every point to the right of *x* on every line connected to *x*; otherwise falsehood.
(3) *F*α is assigned truth at *x* if α is assigned truth at some point or other, to the right of *x*, on each line connected to *x*; otherwise falsehood.
(4) *H*α is assigned truth at *x* if α is assigned truth at all connected points to the left of *x*; otherwise falsehood.
(5) Truth-functions as usual.

This will certainly falsify *CpHFp*, though the pure *GH* portion of the calculus will have the same axioms as before, including *CpHNGNp*.

8. *Propositions that are neither true nor false.* It is a little vexing that no one has yet been able to formalize satisfactorily the ancient and medieval view that predictions of future contingencies are 'neither true nor false'. It is well known that this view provided the original stimulus for Łukasiewicz's 3-valued logic. But that logic has some features which are very counter-intuitive even when we do take the possibility of 'neuter' propositions seriously; in particular, a conjunction of two neuter propositions is neuter, even in the case where one is the negation of the other. If 'There will be a sea-battle' is neuter or undecided, it is no doubt reasonable that 'There will be no sea-battle' should be neuter or undecided too; but not that 'There both will and won't be a sea-battle' should be—that, surely, is plain false. On the other hand, it is equally unplausible to make the conjunction of two neuters automatically false; if they're independent, it is natural that their conjunction should be neuter too. The truth-functional technique seems simply out of place here.

Recently Storrs McCall[1] has attempted to characterize the ancient and medieval position (of which he gives an accurate and well-documented presentation) by means of rules of truth for 'tenseless dated propositions' referring to a time t_0 and asserted at different times. His rules are that

(1) $p(t_0)$ is true at t_0 itself if $p(t_0)$,
(2) it is true at a time earlier than t_0 if there is at that time 'some condition sufficient to make $p(t_0)$ true at t_0',
(3) if $p(t_0)$ is true at any time it is true at all later times,
and
(4) $p(t_0)$ is not true under any other conditions.

An analogous set of conditions is given for falsity, and from his stipulations as a whole it follows that if at any time earlier than t_0 there are not sufficient conditions either to make $p(t_0)$ true at t_0 or to make it false at t_0, then at that earlier time, it is neither true nor false. The conditions are said to be easily adaptable to tensed propositions, but they are so only, so far as I can see, to ones of the form 'It will be (was, is) the case at t_0 that p'. The ancient and medieval view is certainly mirrored with some accuracy in McCall's stipulations; but how they

[1] In 'Temporal Flux', *American Philosophical Quarterly*, Oct. 1966.

work with a detailed linguistic structure, what sort of a calculus we might get (except that we *don't* have to deny $\vdash ApNp$), is left unsaid.

Perhaps 'neither true nor false' is simply a possible way of describing the kind of falsehood which 'It will be that *p*' has, in Peircean logic, when the matter is undecided. It is the actual value we assign to a formula in a Peircean model at points where that formula has different prima-facie values for different routes. In particular, we assign 'neuter' to $K\alpha\beta$, where that, as well as one or both of its parts, has different prima-facie values for different routes; otherwise, as where $\beta = N\alpha$, we assign it falsehood. But how we proceed from there—what use we make of this bit of terminology—I do not know; and I cannot help suspecting that the theory of 'neuter' propositions only arose through a lack of machinery for distinguishing between the two senses of 'will not be', i.e. *NFn* and *FnN*.

Note. Postulates for Peircean metric tense logic (with *Fnp* for the function written *Mnp* on p. 132) may now be found, as part of an improved presentation of metric tense logic generally, in my 'Stratified Metric Tense Logic', *Theoria* 1967.

VIII

TIME AND EXISTENCE

1. *Modalized and tensed predicate logic; the standard systems.* So far we have in a sense considered only tensed *propositional* logic, although we have had quantifiers binding propositional variables and interval variables. We must now consider some of the problems which arise in tensed *predicate* logic, with quantifiers binding individual variables, i.e. variables which (unlike the ones so far used) do stand for genuine names of individual objects.

Here again we have, to begin with, the experience of modal logic to draw upon. One of the principal pioneers in this area was Ruth Barcan Marcus,[1] who took certain Lewis modal systems and appended to them (*a*) some normal postulates for quantification over individual variables, and (*b*) a special 'mixing axiom', $CM\Sigma x\phi x\Sigma xM\phi x$, 'If it could be that something ϕs, then there is something that could ϕ'. This is nowadays often called the 'Barcan formula'. We shall for the moment postpone consideration of the formula's intuitive plausibility, and simply mention one broad feature of the system to which these postulates give rise.

It makes the modal operations behave rather like further quantifiers, 'possibly' resembling 'For some *x*' and 'necessarily' resembling 'For all *x*'. We have, in particular, the following equivalences and implications:

(1) $EL\Pi x\phi x\Pi xL\phi x$, 'Necessarily everything ϕs = Everything necessarily ϕs' (cf. $E\Pi y\Pi x\phi xy\Pi x\Pi y\phi xy$, 'Everything has everything ϕ-ing it = Everything ϕs everything').

(2) $CL\Pi x\phi xM\Pi x\phi x$, 'If necessarily everything ϕs, then possibly everything ϕs'; but not vice versa.

[1] Ruth C. Barcan, 'A Functional Calculus of First Order based on Strict Implication', *Journal of Symbolic Logic*, vol. 11, no. 1 (March 1946), pp. 1–16. Cf. also R. Carnap, 'Modalities and Quantification', ibid., vol. 11, no. 2 (June 1946), pp. 33–64.

(3) $C\Pi xL\phi x\Sigma xL\phi x$, 'If everything necessarily ϕs then something necessarily ϕs'; but not vice versa.

(4) $CM\Pi x\phi x\Pi xM\phi x$, 'If it could be that everything ϕs, then everything could ϕ', but not vice versa (cf. $C\Sigma y\Pi x\phi xy\Pi x\Sigma y\phi xy$, 'If something has everything ϕ-ing it, then everything ϕs something', but not vice versa).

(5) $C\Sigma xL\phi xL\Sigma x\phi x$, 'If something is bound to ϕ, then it is bound to be that something ϕs', though not vice versa.

(6) $C\Pi xM\phi x\Sigma xM\phi x$, but not vice versa.

(7) $CL\Sigma x\phi xM\Sigma x\phi x$, but not vice versa.

(8) $EM\Sigma x\phi x\Sigma xM\phi x$.

Law (1) as well as Law (8) is sometimes also called a (or the) 'Barcan formula'.

It seems a straightforward matter to produce a tensed predicate logic which will have similar laws, and similar warnings (I mean the 'not vice versas'). For example, it was noticed in the Middle Ages that *Semper fuit homo*, 'Always there has been (at least one) man', does not entail that there is at least one man who has existed always, i.e. we do not have $CH\Sigma x\phi x\Sigma xH\phi x$, any more than we have $CL\Sigma x\phi x\Sigma xL\phi x$, the converse of (5) above. Medieval logicians managed, in fact, to be remarkably at home in this area; but what they are most noted for is not the development of a system analogous to the above, but rather their construction of ingenious objections to it, and indeed also to some of the above modal principles themselves. Buridan, for example, objected to (4) that it could be that everything is God ($M\Pi x\phi x$), and that this actually was the case before the creation ($P\Pi x\phi x$), and would be the case if God were to annihilate all other beings; but it is just not true that everything could be God ($\Pi xM\phi x$), or that everything has been God ($\Pi xP\phi x$)—most of us neither have been nor could be.[1]

2. *Ancient, medieval, and modern objections to coming to be, being brought into being, and being prevented from being.* Implicit objections to a tensed predicate logic of this sort are to be found not only in technical logical works but in general philosophical discussions of the concept of coming to be, and this not only in the medieval period but in the ancient and modern periods also.

[1] *Sophismata*, ch. 4, *sophisma* 13.

There is, for example, the following argument recorded in Aristotles *Physics*.[1]

> 'The first of those who study science . . . say that none of the things that are . . . comes to be . . . , because what comes to be must do so either from what is or from what is not, both of which are impossible. For what is cannot come to be (because it *is* already), and from what is not nothing could have come to be (because something must be present as a substratum).'

I think this argument might be summed up in the following diagram:

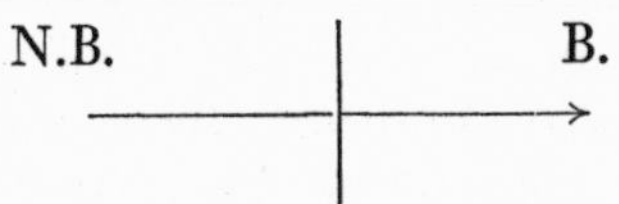

Here the left-hand compartment represents the realm of non-being, the right-hand compartment the realm of being, and the arrow the path of something that is supposed to come to be. But if the left-hand compartment really does represent the realm of non-being, the portion of the arrow on that side of the line has no business to be there—on that side of the line there just isn't anything to carry out this part of the performance. That leaves the right-hand compartment, but whatever it is that is going on there, it cannot be 'coming to be', for what's in *that* compartment is what already *is*. The argument seems to me conclusive, though it should be noted that however it tells against the conception of *coming* to be, it does not make it impossible for a thing to *start* to be, i.e. to exist for the first time —this takes place unambiguously on the right-hand side, and at least as far as this argument goes, there is no reason why such things should *not* take place there. But this line of argument tells strongly against a formula which would be easily obtainable if we appended ordinary laws of quantification theory to most of the tense-logics we have been considering, namely $C\Sigma x\phi xP\Sigma xF\phi x$, 'If something is ϕ-ing (e.g. existing), then there used to be something that was going to be ϕ-ing.'

A very similar argument was mentioned by Thomas Aquinas as a possible objection to the doctrine of creation out of nothing.[2]

[1] 191ª 23–32.

[2] Aquinas, *De Potentia Dei*, Q. 3, Art. 1, Obj. 17. The philosophical importance

The objection runs:

> 'The maker gives being to that which is made. If then God makes a thing out of nothing, he gives being to that thing. Hence either there is something that receives being, or there is nothing. If nothing, then nothing receives being by that action of God's, and thus nothing is made thereby. If something, . . . God makes a thing from something already existing, and not from nothing.'

The concept of being *brought* or 'launched' into existence has the same difficulties as that of 'coming' into existence. The same diagram will do for this as for the preceding; the only difference is that the object is supposed to be *helped* over the fence between non-being and being; and once again, if the starting point really *is* non-being, there is just nothing there to be helped; and if not, it is not *existence* that the thing is being helped to, since it already has it. With problems of this sort in mind, P. T. Geach[1] has suggested that bringing a man into being out of something, i.e. making something a man, may be reported by the form:

(1) For some x (God has brought it about that (x is a man));

while making a man out of nothing may be reported by:

(2) It is not the case that (i); but God has brought it about that (for some x (x is a man)).

The fact that the second part of (ii) does not imply (i) means that there is no Barcan formula for 'bringing it about that'. It may be noted that we also have a distinction here like that made in Buridan's *sophisma* of the man who says 'I promise to give you a horse'.[2] I cannot actually *give* you a horse without there being some horse that I give you, but I *can* promise to give you a horse without there being any particular horse that I promise you. Similarly, what God does in creating a man out of nothing, on Geach's account of it, is not to say 'Let *this* man be', and then this man is, but rather to say 'Let *a* man be' (maybe a man with such-and-such further detailed specifications) and then this man is. There is no 'this' until the man is already there.

of Aquinas discussions of this subject was first brought home to me by A. Sertillanges, *L'Idée de Création et ses Retentissements en Philosophie*.

[1] P. T. Geach, 'Causality and Creation', *Sophia* (Melbourne), vol. 1, no. 1 (April 1962), pp. 1–8.

[2] *Sophismata*, ch. 4, *sophisma* 15.

An allied point emerges from a passage in Jonathan Edwards,[1] where he is maintaining that God has no freedom of choice since a perfect being will always choose the best possible course. To the objection that God can at least decide either way if the choice is morally indifferent, Edwards replies that no choices are. But it might be said that if God has placed two exactly similar objects, at their creation, in different places, it could have made no moral difference if he had placed them the other way round. Edwards's first answer to this is that if the objects really differ in nothing but their position, there is no difference between the two alleged alternatives.

He recognizes, however, that it might be said that the objects (he makes them two spheres) are supposed *numerically* different, so that there *is* a difference between *A* being at *X* and *B* at *Y*, and the opposite placing of them. His reply to this is obscure, but it suggests that even if *A*'s being at *X* and *B* at *Y is* different from the other placing, God's *creating A* at *X* and *B* at *Y* couldn't have been a different divine decision from creating *B* at *X* and *A* at *Y*. For if it were, all sorts of other choices might also have confronted him, of a kind which are clearly ridiculous.

> 'If, in the instance of the two spheres, perfectly alike, it be supposed possible that God might have made them in a contrary position; that which is made at the right hand, being made at the left; then I ask, whether it is not evidently equally possible, if God had made but one of them, and that in the place of the right-hand globe, that he might have made that numerically different from what it is, and numerically different from what he did make it; though perfectly alike, and in the same place . . . ? Namely, whether he might not have made it numerically the same with that which he has now made at the left hand, and so have left that which is now created at the right hand, in a state of non-existence? And if so, whether it would not have been possible to have made one in that place, perfectly like these, and yet numerically different from both? And let it be considered, whether from this notion of a numerical difference in bodies, perfectly alike, . . . it will not follow, that there is an infinite number of numerically different possible bodies, perfectly

[1] Op. cit., Part IV, Section viii.

alike, among which God chooses, by a self-determining power, when he sets about to make bodies.'

This conclusion does *not* follow from that 'notion', but it *does* follow, or something like it does, from the notion that individuals have a distinct identity *before they exist*. If God can say 'Let *this* go here, and *that* go there' of things that do not yet exist but will do so, there is nothing to stop him from making decisions about what not only does not yet exist but will never do so. And indeed the supposition of qualitative likeness in the end-products is a superfluous circumstance; if God can say even 'Let *this* be a perfect sphere and *that* a dented one' of things that do not yet exist but will do so, there is again nothing to stop him from making decisions about what not only does not yet exist but never will ('Let *that* just stay as it is'). Edwards takes this to be obviously absurd, and the same would be true, I suggest, of similarly particularized *prophecies*. Suppose some gifted gipsy or Cornish man to go into a trance in 1850 and say 'Next century there will be a person called A. B. with such-and-such a character and history, and a person called M. N. with such-and-such a different character and history'; and then suppose the man suddenly to get worried and say 'No, perhaps it's the *second* man I meant who is going to be called A. B. and have the first character and history, and the first who will be called M. N. and have the second', and then he gets still more worried and says 'Perhaps I am even more wrong than that, and it is neither of the persons I meant who will do and suffer these things, but two quite different individuals altogether'. These worries are surely senseless, and the alternatives *at that time* not distinct.[1]

Ryle, in our own period, has made a similar point to Thomas's, not about the conferring but about the *prevention* of existence:

> 'If my parents had never met, I should never have been born . . . So we want to say that certain circumstances would have prevented me from being born . . . But then there would have been no Gilbert Ryle . . . for historians to describe as not having been born . . . What does not exist . . . cannot be named, individually indicated or put on a list,

[1] Cf. A. N. Prior, 'Identifiable Individuals', *Review of Metaphysics*, Dec. 1960.

and cannot therefore be characterised as having been prevented from existing.'

The same diagram does for this as for the others, except that one pictures the object's progress, while in the realm of non-being, as meeting with an obstruction, thus:

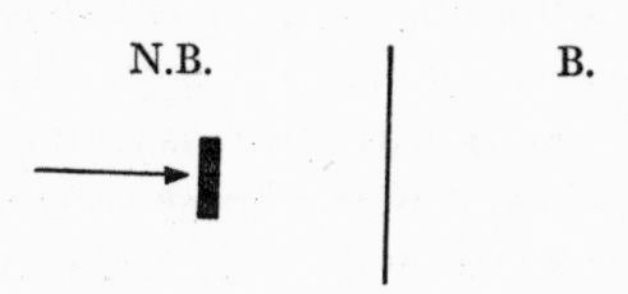

But once again it is absurd to suppose anything of this sort going on *there*, while if one transfers the arrow and the obstruction to the other side, it cannot represent something being prevented from starting to exist, but only something that already exists being destroyed. Ryle goes on:

> 'This point seems to me to bring out an important difference between anterior truths and posterior truths, or between prophecies and chronicles. . . . After 1900 there could be true and false statements . . . mentioning me. But before . . . 1900 there could not be true or false statements giving individual mention to me. . . . While it is still an askable question whether my parents are going to have a fourth son, one cannot use the name "Gilbert Ryle" or use as a pronoun designating their fourth son the pronoun "he". Roughly, statements in the future tense cannot convey singular, but only general truths.' [1]

(Only roughly, because they might convey truths about the future feats of already existing individuals.) 'It will be that someone is the Ryles' fourth son' does not entail 'It is true of someone that *he* will be the Ryles' fourth son' $(\dashv CF\Sigma x\phi x\Sigma xF\phi x)$.

3. *Ampliation.* There are certain movements of quantifiers inside and outside other operators which look as if they would be easy, but which in cases like the preceding have encountered

[1] G. Ryle 'Dilemmas', pp. 25–27. The same topic is nicely handled by Michael Frayn in 'The men who never were', *Observer*, 27 Feb., 1966, p. 10.

obstacles; and the immediate source of these obstacles is obvious enough. When a quantifier is governed by, say, a tense operator, it is natural to think of it as ranging over such objects as there may be at the time to which the tense operator takes us; for example, 'It will be that something ϕs' is most naturally read as 'It will be, at some future time, that something *then* existing ϕs'. On the other hand, a quantifier *preceding* any such operator is naturally taken to be governed by the 'It is the case that—' which is prefixable to anything we say, and therefore to range over what *now* exists. And where these ranges do not coincide—as is bound to be the case where we are considering what now is but once was not, or (in the case of modal logic) what in fact is, but need not have been—we have to tread carefully.

Medieval logicians had considerable sensitivity to problems of this sort, but in their solutions to them were hampered by an inadequate analysis of quantifiers and tenses. They mostly handled propositions like 'Some man will be running' in which the sign of quantity was attached to some specific common noun and the sign of tense to a following verb. They held that a noun like 'man' normally stands for (*supponit pro*) presently existing men, in the sense that any presently existing man's ϕ-ing, and only a presently existing man's ϕ-ing, will verify 'Some man ϕs'. But where the verb is tensed, and in some other circumstances, the *suppositio* of the subject-noun will be widened or 'ampliated' to include also objects to which it *was* applicable, or to which it *will be* (depending on the tense of the verb). 'Some man will be running', for example, would be verified by a man's running in the future, even if that man doesn't exist yet. This ruling had some odd consequences; Buridan[1] was compelled to agree, for example, to *Senex erit puer*, 'An old man will be a boy', on the grounds that this means that someone who *is or will be* an old man (e.g. someone who is now a baby, or unborn) will be a boy. To give the sense of 'No old man will be a boy' in which that is true, one has to say explicitly 'Nothing that *now is* an old man will be a boy'. But many knots were untied this way, e.g. they could say that 'Some house doesn't exist' is false although it appears to follow from 'Nothing that has perished exists, and some house has perished';

[1] *Sophismata,* ch. 4, *sophisma* 4.

for the minor in this means 'Some *present or past* house has perished' while the conclusion means 'Some *present* house doesn't exist'.

With tense-operators and quantifiers both prefixed to open sentences, we can let the range of the quantifier be settled by the order in which the different prefixes go. If we do this, however, we may have to do more than be careful in *this* part of our tense-logic; there could be repercussions in the propositional part of it too. In developing this point, we may start from an objection that was raised against one of the 'Barcan formulae' a few years ago by John Myhill.[1]

4. *Objections to standard modal logic suggested by Myhill, Ramsey, and Chrysippus.* Myhill, in discussing the formula $C\Pi xL\phi xL\Pi x\phi x$, starts from the assumption that not only *which* objects the universe contains, but also *how many* of them there are, must be a contingent matter. Suppose there are in fact five—*a*, *b*, *c*, *d*, and *e*. Then *a* is necessarily identical with *a* (everything is necessarily identical with itself), and so is necessarily either identical with *a* or identical with *b* or identical with *c* or identical with *d* or identical with *e* (*CLpLApq*). Similarly, *b* is necessarily identical with *b*, and therefore with *a* or *b* or *c* or *d* or *e*. They are *all*, in fact, each for its own reason, necessarily either identical with the 1st or with the 2nd or with the 3rd, etc. If we let this necessary disjunction of identities be ϕ, we have here $\Pi xL\phi x$. But there didn't have to be just 5 individuals, so it didn't have to be true that everything is either *a* or *b* or *c* or *d* or *e*, i.e. we *don't* have $L\Pi x\phi x$; and so we have a counter-example to the Barcan formula. Or, if the Barcan formula *is* true, there aren't just 5 individuals, for if there were the Barcan formula would lead us from that to a falsehood. Nor, by similar reasoning, can *any* finite number *n* be the number of individuals, if the Barcan formula is true; so if it is true the number of individuals would be infinite. But drawing this conclusion also would make the number of individuals in the universe a logical matter. So the Barcan formula *can't* be true, and must be dropped.

If it can be, that is. Myhill's description of what he is

[1] J. Myhill, 'Problems arising in the Formalisation of Intensional Logic', *Logique et Analyse*, April 1958, pp. 76–83.

dropping it *from* is not very clear. He says that the system he is after consists of a normal basis for predicate calculus together with 'the' axioms for Lewis's S5, the 'rule of necessitation' (to infer $\vdash L\alpha$ from $\vdash\alpha$), and nothing else; above all not *CΠxLφxLΠxφx*. But if by 'the' axioms for S5 he means Lewis's own original ones, these all start in effect with *LC*, and are geared to rather special rules; neither *C*-detachment nor the rule of necessitation will get any theorems from them at all (except the axioms, and substitutions in them, preceded by *L*'s). He almost certainly means Gödel's formalization, with *CLpp*, *CLCpqCLpLq*, and *CNLpLNLp* subjoined to propositional calculus, or to predicate calculus, with RL. But if this is what he means, he is not in a position just to take or leave the Barcan formula, since on that basis it is provable.[1]

In the original Barcan basis, indeed, the equivalent formula was an independent axiom, and so could be dropped if desired, because the modal system used was not S5 but a weaker one. One might, therefore, consider meeting Myhill's problem by weakening his modal logic to, say, S4. That quantified S4, without special additions, does not contain *CΠxLφxLΠxφx*, was shown by Lemmon in 1960.[2] Lemmon has also shown, however, in 1965, that this formula *is* provable in the quantified 'Brouwersche' system, i.e. T+*CpLMp*, or *CMLpp*, as follows:

1. *CMΠxLφxMLφx*	(*CΠxψxψx*, *ψ/Lφ*; RMC)
2. *CMΠxLφxφx*	(1, *CMLpp*)
3. *CMΠxLφxΠxφx*	(2, *Π*2*x*)
4. *CLMΠxLφxLΠxφx*	(3, RLC)
5. *CΠxLφxLΠxφx*	(4, *CpLMp*).

We have already seen that the system B is what we get for $L\alpha = KK\alpha G\alpha H\alpha$ even when for *G* and *H* we use the 'minimal' tense-logic K_t. So, however helpful or plausible it may be to dismiss the Barcan formula from *modal* logic by working from S4 (or something weaker) instead of S5, this move doesn't look

[1] For the proof, from this basis, of the equivalent formula *CMΣxφxΣxMφx*, see A. N. Prior, 'Modality and Quantification in S5', *Journal of Symbolic Logic*, vol. 21, no. 1 (March 1956), pp. 60–62.

[2] E. J. Lemmon, abstract in *Journal of Symbolic Logic*, vol. 25, no. 4 (Dec. 1960), pp. 391–2.

as if it will work in tense-logic. Indeed, the above proof only needs slight modification to obtain, in K_t, a Barcan formula for G; for we have the following:

1. $CP\Pi xG\phi xPG\phi x$ $(C\Pi x\psi x\psi x,\ \psi/G\phi;\ \text{RPC})$
2. $CP\Pi xG\phi x\phi x$ $(1,\ CPGpp)$
3. $CP\Pi xG\phi x\Pi x\phi x$ $(2,\ \Pi 2x)$
4. $CGP\Pi xG\phi xG\Pi x\phi x$ $(3,\ \text{RGC})$
5. $C\Pi xG\phi xG\Pi x\phi x$ $(4,\ CpGPp)$.

A Barcan formula for H may be proved analogously, and appropriate ones for F and P by contraposition from these. (Cocchiarella had in 1965 a similar direct proof of $CP\Sigma x\phi x\Sigma xP\phi x$ and its F-image in which, although his system is stronger than K_t, he used only theses which are in fact in K_t.) And even in modal logic as ordinarily interpreted there is abundant evidence that something more radical is needed than the weakening of S5 to one of the other standard systems.

Only very weak modal assumptions are made in an argument quite like the first part of Myhill's which F. P. Ramsey put forward over thirty years before.[1] Ramsey did not hold that 'no proposition concerning the cardinality of the universe (except the one asserting its non-emptiness) is necessary'; on the contrary, he believed that any such proposition would be either a tautology or a contradiction—either necessary or impossible. He adopted the view of Wittgenstein's *Tractatus* that 'For all *x*, *ϕx*' is just short for the long conjunction '*ϕa* and *ϕb* and *ϕc* . . .', and that what the latter form apparently needs to have added before it can yield the former, namely '*a*, *b*, *c*, . . . are all the individuals', is, when true, logically necessary. Similarly propositions of the form '*a*, *b*, *c* . . . are *not* all the individuals' are, when true, necessary. Those who object to this, he says, will surely admit that (1) 'numerical difference and identity are necessary relations', that (2) 'There is an *x* such that "*fx*" follows from "*fa*"' and that (3) 'whatever follows necessarily from a necessary truth is itself necessary'. Suppose now that the universe in fact contains not only the objects *a*, *b*, and *c* but a further object *d*. By (1) it will be a necessary truth that *d* is not identical either with *a* or with *b* or with *c*,

[1] In 'Facts and Propositions', *Proc. Arist. Soc.*, supp. vol. 8 (1927), reproduced in *The Foundation of Mathematics*.

and by (2) and (3) it follows from this that it is a necessary truth that there is something that is neither *a* or *b* or *c*, i.e. that these are not all the individuals.

It should be noticed that Ramsey does *not* first say that because *d* is necessarily other than *a*, *b*, and *c*, therefore *there is something that is necessarily* other than them, i.e. he does not argue from *Lφd* to *ΣxLφx*, and then from this to '*Necessarily there is something* that is other than them', *LΣxφx*, using one of the formulae from the Barcan calculus of which we have learnt to be suspicious. He just uses *CφdΣxφx*, *CLCpqCLpLq*, and he gets his *Lφd* (and so *LΣxφx* from the other two) from the assumption that 'numerical difference and identity are necessary relations'. This assumption has certainly been much criticized in recent years, but he wrote in 1927, and the assumption was hardly ever questioned until Mrs. Marcus proved it (or at least proved it for identity) ten years later. Remember that he didn't mean by it anything like 'The Morning Star is necessarily identical with the Evening Star'; he operated with Russellian proper names, and his *CIxyLIxy* meant simply that each thing cannot but be that individual thing that it is (what would it be for *it* to be something else?), and his *CNIxyLNIxy* that nothing *can* be another thing. This *isn't* quite so obvious as he and his contemporaries thought. But it is certainly not the only premiss of Ramsey's that can be questioned, if we do not like his conclusion.

One of the others, *CLCpqCLpLq*, was long ago questioned by Chrysippus. This law is one which occurs in Aristotle with variations; he says not only that what necessarily follows from the necessary is itself necessary, but also that what necessarily follows from what is possible is itself possible (*CLCpqCMpMq*), and that the impossible does not follow from the possible. It was this last form with which Chrysippus was most concerned. It was, as we have seen, a premiss of the Diodorean 'Master argument', but although it is mentioned in connexion with that argument that Chrysippus did not accept this premiss, the only near-detailed account that we have of why he did not accept it has nothing to do with Diodorean definitions of possibility, but has rather to do with worries about non-existence. He is said to have argued that 'If Dion is dead, this man is dead', uttered when Dion is being indicated, is a 'sound conditional' in which the consequent follows from the antecedent, but that while it is possible for Dion to be dead, 'this man is dead' could never be true, since if Dion did not exist there would be no such proposition as the one that it now expresses.[1] This argument has not in general struck historians of logic as impressive, and I too find it a little unconvincing because of the obscurity of the sense in which ϕ (Dion) is supposed to entail ϕ (this man). I believe, however, that very little alteration of it does produce a reason for denying *CLCpqCMpMq*, or anyhow for denying *CNMNCpqCMpMq*, and also for denying the tense-logical *CNPNCpqCPpPq*.

Before making this amendment, however, a little should be said in defence of Chrysippus's contention that under certain circumstances we would not only not be able to express certain propositions which we now can, but there would be no such propositions; and the analogous view in tense-logic that there have been times at which not only were men not able to express certain propositions which they now can, but there *were* no such propositions. This view is in a way already implicit in the comments that have been made above on coming to be, being brought into being, and being prevented from being, and especially in Ryle's discussion of the last. But its justification will be clearer if we look at one more philosopher's discussion of existence, modality, and time; only in this example quantification will definitely not be involved, so that there can be no suggestion that all our troubles are with that.

5. *Moore on what might not have existed, and on what once did not exist.* Russell has often said that it does not make sense to attach 'exists' or 'does not exist' to what he calls a logical proper name, i.e. an expression whose function in a sentence is purely to indicate which object we are talking about, and not to describe the object in any way. We can attach 'exists' or 'does not exist' to a description, e.g. 'The man on the moon exists' and then the predicate is eliminable by certain well-known means. But '*This* exists', '*This* does not exist' are senseless. This, however, has been questioned by Moore, and it seems to me that Moore

[1] See W. C. Kneale and Martha Kneale, *The Development of Logic*, p. 126; and M. Kneale, 'Logical and Metaphysical Necessity', *Proc. Arist. Soc.* 1937–8, pp. 253–68.

at this point propounds a view which fits much better than Russell's own view does into Russell's general logical position. What Moore suggests[1] is that 'This exists' and 'This does not exist' are not necessarily senseless, but may be so used that *if* they are not senseless, the former is bound to be true and the latter false. For if the function of 'This' in a sentence is purely to indicate the object the sentence is about, then if in fact no object is indicated, no sentence containing *this* 'This' really says anything, and of course 'This exists' and 'This doesn't exist' fall with the rest. But if 'This' does pick out the object intended, what 'This exists' says will have to be the case and what 'This doesn't exist' says, cannot be. It may be noted that although Russell rejects 'This exists' as ill-formed, the form 'x is identical with x' as used in *Principia Mathematica* has exactly the properties that are ascribed to 'This exists' by Moore, and could be used to define it.

One reason Moore gives for believing that 'This exists' can have a sense at least of this sort, is that 'This might not have existed' is something which is certainly not without meaning and which is in general true. The bearing of this fact on the main argument is, I think, that a compound sentence cannot be meaningful if a component sentence in it is not, and 'This exists' is a component out of which 'This might not have existed' is constructed. The construction is presumably 'It could have been that (it is not the case that (this exists))', $MNE!x$ (using '$E!x$' for 'x exists'). But if this *is* the construction, what is said is surely not true. For Moore himself says that 'This doesn't exist', i.e. 'It is not the case that this exists', is not true under any circumstances in which it says anything, and so far as I can see it never could be; so $MNE!x$ is bound to be false. But there *is* a sense of 'This might not have existed' in which what it says could be the case (and generally is), i.e. the sense: 'It is not the case that (it is necessary that (x exists))' $NLE!x$. There are, then, no possible states of affairs in which it is the case that $NE!x$, and yet not all possible states of affairs are ones in which $E!x$. For there are possible states of affairs in which there are no

[1] G. E. Moore, 'Is Existence a Predicate?', *Proc. Arist. Soc.* supp. vol. 15 (1936), reproduced in *Philosophical Papers*. The same points are developed in Moore's *Lectures on Philosophy* (1966), p. 40, and above all in the quite perfect little piece on 'Necessity' (from lectures of 1925–6) on pp. 129–31.

facts about x at all; and I don't mean ones in which *it is the case that there are not* facts about x (for that would itself be one, if true), but ones such that *it isn't the case in them that there are* facts about x.

The relation of existence to time is similar, and Moore was as clear about this as he was about possibility.[1] 'I don't exist now' and 'This doesn't exist now', he says, 'are self-contradictory'. But '"I might not have existed now (at t_1)" or "This mightn't" are not, because what they mean is merely that there would have been no contradiction in my saying of myself in the past "I shan't exist at t_1", and will be no contradiction in my saying of myself in the future "I didn't exist at t_1"'. He is then, however, careful to add, 'No-one could, of course, have said of "this" in the past "this won't exist at t_1", unless this did exist at the past moment in question; nor could anyone say of "this" in the future "this didn't exist at t_1", unless "this" exists at the future moment in question.' It is clear from this that if someone says *truly* 'I didn't exist at t_1', the truth of this cannot consist in there having been a fact at t_1, which someone could have expressed by then saying 'This doesn't exist', since that is *always* 'self-contradictory'; i.e. it doesn't mean 'It was the case at t_1, that (I don't exist)'; it can only mean 'It was not the case at t_1, that (I exist)', i.e. it *now is not* the case that my existence was the case then—it's not that my *non*-existence *then was* the case.

All this just follows from the rubbing out of the line in the left-hand compartment of the diagram about coming-to-be. There are just no facts at all in that compartment. And one thing should now be said about the tense-logical law which I said this rubbing-out falsified: $C\Sigma x\phi xP\Sigma xF\phi x$, 'If something ϕs (e.g. exists) then it was the case that something was going to ϕ (e.g. exist)'. The quantifiers can go from this; I mean, it still has to be denied if you leave it at $C\phi xPF\phi x$, 'If *this* exists, it has been going to exist'; or indeed if you leave it at $CpPFp$.

6. *Arguments against some common principles of modal and tense-logic.* We can now return to Chrysippus and Ramsey, and the laws $CLCpqCLpLq$ and $CLCpqCMpMq$. Since $NL \neq MN$ (we have, e.g. sometimes $NLE!x$ but never $MNE!x$), we cannot simply equate L, 'true in all possible states of affairs', with NMN, 'false in none'; and we need to consider whether the L in these laws

[1] *The Commonplace Book of G. E. Moore*, p. 329; cf. also pp. 236–7.

really means 'true in all' or just an abridgement of 'false in none'. Taken in the former sense, the laws are true, but of very limited application (what *is* 'true in all states of affairs'?); in the latter sense, they are not true. Consider *CNMNCpqCMpMq* first, with a Chrysippus-like example. 'This man doesn't exist', we may agree with him is in no circumstances true, where the 'this' is supposed to identify an individual, though in some circumstances there may be no such proposition as the one that it now expresses. That is, we don't have *MNE*!*a*, though we don't have *LE*!*a* either. And 'If nothing exists, this man doesn't' is never false (*NMNCNΣxE*!*xNE*!*a*), for it is true whenever there is such a proposition. And it *is* possible that nothing should exist, *MNΣxE*!*x*. So here we have an *NMNCαβ* and an *Mα* which are true, though the corresponding *Mβ*, namely *MNE*!*a*, is false; i.e. *CNMNCpqCMpMq* does not universally hold. And the example is *almost* Chrysippus's own, except that I have replaced his 'Dion doesn't exist' by 'Nothing exists', the entailment by which of 'This man doesn't' is perhaps clearer. It is perhaps a little contentious to say that it could be that nothing exists, but if one held that being of the basic *sort* one is, e.g. being a man, is 'essential' or 'necessary' in anything that *is* of that sort, one could say that it could not be false that if no *man* exists then this man doesn't, that it could be that no man exists, and that it couldn't be (*isn't* the case in any possible state of affairs) that precisely *this* man doesn't exist.

We can deal similarly with Ramsey and his world of four individuals. Where *φd* is '*d* is neither *a* nor *b* nor *c*' we do have, I think, *NMNCφdΣxφx*, 'It could not be false that if *d* is neither *a* nor *b* nor *c* then something is neither *a* nor *b* nor *c*'. We also have *NMNφd*, 'It could not be false that *d* is neither *a* nor *b* nor *c*', though there *could* just be no such proposition as this one, and would be if any one of *a*, *b*, *c*, or *d*, were non-existent. But *NMNΣxφx*, 'It could not be false that (something is neither *a* nor *b* nor *c*)', is *not* true, for this would be false if *d* didn't exist (really false, and not itself non-existent, since it doesn't mention *d*). So we don't have here *CLCpqCLpLq* in the sense of *CNMNCpqCNMNpNMNq*.

In tense-logic, counter-examples to *CNPNCpqCPpPq* ('If it has never been false that if *p* then *q*, then if it has been that *p*, it has been that *q*') are easier to construct. To falsify it we need only find some object x which has been in existence longer than some other object y, some ϕ which was true of x before y existed but has not been since, and some ψ which has never been true of y at all, and let our p be ϕx and our q be $A\phi x\psi y$. One could, e.g. adapt a modal example used in *Time and Modality*[1] with God for x and me for y, 'God alone exists' for ϕx, and 'I don't exist' for ψy (it has never been false, though it has sometimes been unstatable, that if God alone exists then either God alone exists or I don't exist; it has been the case—on the Christian hypothesis—that God alone exists; but since 'Either God alone exists or I don't exist' has been statable it has never been true). But there is no need to bring God or existence into it. For example, we could use 'That' to indicate a small child who has never, among other things, driven a Cadillac, and 'this' to indicate an older person who went to school before this child was born but hasn't done so since. We then have

(1) It has never been false that if *this* person is going to school then either *this* person is going to school or *that* person is driving a Cadillac ($NPNC\phi xA\phi x\psi y$).

(There was, indeed, no such proposition as this before that person existed, but the proposition has never been *false*.) We also have

(2) It has been the case that *this* person is going to school ($P\phi x$).

On the other hand we don't have

(3) It has been the case that: either *this* person is going to school or *that* person is driving a Cadillac ($PA\phi x\psi y$).

For since *that* person started to exist both parts of the disjunction have been false, and so the whole disjunction false, and *before* that person existed there was no such proposition as the last disjunct ('*that* person'—meaning the one we mean now—'is driving a Cadillac'), and so no such proposition as the disjunction; which disjunction, therefore, expresses something that has *never* been the case, falsifying (3) and so falsifying C(1)C(2)(3).

[1] p. 49.

7. The modal system Q, its modifications, and its adaptation to tense logic. In *Time and Modality* there is adumbrated a modal system called Q, intended as a reasonably strong modal logic which would nevertheless lack such dubious principles as *CNLpMNp* and *CNMNCpqCMpMq*, and which could be combined with a normal quantification theory without yielding the dubious principles in the mixed field that have been mentioned earlier. Q was not axiomatized but was characterized by a matrix, the possible values of propositions being infinite sequences of 1, 2, and/or 0, the first member never being a 2. A 2 at a point in a sequence meant that there is no such proposition as the one in question in the world represented by that point. *All* compounds have 2s at *any* places where *any* of their components have 2s (where there is no such proposition as *p*, there are no functions of *p* either). Otherwise, the *Np* sequence interchanges the 1s and 0s of the *p*-sequence; the *Kpq* sequence has 1s where both the *p*-sequence and the *q*-sequence do; otherwise 0s; *Mp* has 1s everywhere (always apart from where the 2s are) if *p* has 1s anywhere; *Lp* has 0s everywhere (apart from the 2s) unless *p* has 1s everywhere (*every*where—no 1s in *Lp* if *p* has 2s). A formula is a law if its sequence never has 0s in it for any values of its variables.

No set of postulates was then known for which this matrix was characteristic, but in a paper published in 1964,[1] R. A. Bull proved completeness for a set which took as undefined my own strong *L* ('true in all worlds') and a weaker *L* equivalent to my *NMN* ('false in none'). As a corollary to this result, it was possible to prove completeness for some simpler postulates which I had put forward tentatively in 1959, taking as undefined my original *M* and a function *Sp*, suggested by J. L. Mackie, which could be read as 'always statable' and was equivalent to *LCpp* (strong *L*).[2] My original *Lp* could then be defined as *KSpNMNp*, '*p* always statable and never false'. The postulates, subjoined to propositional calculus with substitution and detachment, were as follows:

RS1 : ⊢*CSαSp*, where *p* is any variable in α;

[1] R. A. Bull, 'The Axiomatisation of Prior's Modal Calculus Q', *Notre Dame Journal of Formal Logic*, vol. 5, no. 3 (July 1964), pp. 211–14.

[2] A. N. Prior, 'Notes on a Group of New Modal Systems', *Logique et Analyse*, April 1959, pp. 122–7.

RS2 $\vdash CSpCSqCSr \ldots S\alpha$, where p, q, r, ... are all the variables in α;

RSM: $\vdash C\alpha\beta \rightarrow \vdash CSpCsq \ldots CM\alpha\beta$, where β is fully modalized (i.e. all its variables within the scope of an S or M) and p, q ... are all the variables in β that are not in a;

and the axiom *CpMp*. If $\vdash Sp$ is added to this, *Lp* collapses to *NMNp* and the system becomes S5 (this amounts to removing the possibility of propositions just not figuring in certain worlds; or removing all sequences with 2s from the matrix).

I presented Q as a 'logic for contingent beings'; meaning by that a logic in which one could intelligibly say that some beings are contingent and some necessary. Lemmon pointed out that a real 'logic for contingent beings' would exclude the second group, and one would get it by deleting from Q's matrix all sequences *not* containing 2s, and perhaps axiomatize it by adding $\vdash NSp$ to Q's postulates. Lemmon also noticed two other possible modifications of Q. In one, we delete from Q's matrix all sequences which contain both 1s and 0s but not 2s; a possible axiomatization is by adding $\vdash CSpCMpp$ to Q. Here, as in Q, there is room for both necessary and contingent beings, but all truths which are *purely* about necessary beings (and therefore always statable—have no 2s in their sequences) are themselves either necessary or impossible; though ones which are about *both* necessary and contingent beings—e.g., perhaps, '9 is the number of the planets'—may be contingent. Finally, we may delete from Q's matrix all sequences whatever that have both 1s and 0s; and add $\vdash CMpp$ to the postulates. This makes $Mp = p$ and $Lp = KSpp$, 'necessarily statable and actually true'. If we call a proposition 'pure' if it contains no references to particular contingent beings, and 'impure' if it has such references (even if what it says of such beings is just, e.g. that if they're red they're red), the 'necessary' truths and falsehoods of this last system are the 'pure' truths and falsehoods, and the 'contingent' ones the 'impure' ones. The matrix for this is equivalent to a 4-valued one.

A modification of tense-logic analogous to Q has still to be attempted, though Q can of course be taken over as it stands with *L* for the temporal 'always', *M* for 'sometimes', *NMN* for 'never not' and *NLN* for 'not always'. One or two details of the presupposed $GPHF$ calculus are obvious. For example, we do have $CHCpqCHpHq$ but don't have $CNPNCpqCPpPq$ (H and P are not interdefinable); and we don't have the rule RH, to infer $\vdash Ha$ from $\vdash a$. For example we have $\vdash C\Pi x\phi x\phi y$, 'If everything ϕs then y ϕs', this being true for any y of which the formula can say anything, i.e. any y there *is*; but we don't have $\vdash HC\Pi x\phi x\phi y$, 'It has always been that if everything ϕs then y does', for even $C\Pi x\phi x\phi y$'s which are true now were not true—or anything else—before y existed. Whether we should have a rule to infer the weaker $\vdash NPN\alpha$ from $\vdash\alpha$, is a tricky question. Ordinary quantification theory and identity theory gives us $\vdash \Sigma xIxx$, 'Something is itself', which we can equate with 'Something exists', and this with the proposed rule would give us $\vdash NPN\Sigma xIxx$. 'It has never been false that (something exists)', i.e. it has never been the case that nothing exists. We can deal with this problem either (1) by having some non-standard quantification theory with identity in which $\Sigma xIxx$ is not provable, or (2) by denying the rule to infer $\vdash NPN\alpha$ from $\vdash\alpha$, or (3) justifying $\vdash NPN\Sigma xIxx$, e.g. on the grounds that before anything existed there was no such proposition, and therefore no such true proposition, as $N\Sigma xIxx$. This last may sound even trivially right—if there's nothing how can there be propositions?—but propositional 'existence' is not to be taken as literally as that: it is a being-the-case-or-not rather than a literal being; so that bit of universal instantiation won't do. Nor is $N\Sigma xIxx$ directly *about* any individual in the way that $NIaa$ would be, so we can't argue that there would be no such proposition in an empty universe because there would be no such object there as the one that it is about.

The correct answer, i.e. the answer which is in accordance with the intuitions behind this sort of system, seems to me to be as follows: We dismiss solution (2), on the grounds that the rule to infer $\vdash NPN\alpha$ from $\vdash\alpha$ simply reflects the fact that what is meant by calling a formula a 'theorem' of this system is that any constants that we put for its free variables will give us something which is never false. We can then either accept $NPN\Sigma xIxx$ as a theorem in its normal sense, as meaning that the universe has never in fact been empty, or if we do not wish to commit ourselves on this point, reject $\Sigma xIxx$ as a theorem, i.e. as something we commit ourselves to as being never false. If we take this last line, however, we shall have to modify the rule of detachment, since both *Iaa* and *CIaaΣxIxx* are certainly theorems in the above sense, but *ΣxIxx* is not. A form of quantification theory with just this peculiarity, designed to cope with the possibility of empty universes, was put forward by Mostowski in 1951.[1] Mostowski modifies detachment to:

If all individual variables free in α occur free in β, then if $\vdash\alpha$ and $\vdash C\alpha\beta$ then $\vdash\beta$.

In other respects his quantification theory is normal; e.g. it has both $\vdash C\phi y\Sigma x\phi x$ and $\vdash C\Pi x\phi x\phi y$, and the rule to infer $\vdash \Pi xa$ from $\vdash\alpha$. This complication also affects the analogous modal system Q, and in *Time and Modality*[2] I did foresee trouble with detachment in extensions of Q, though I was over-optimistic about the possibility of retaining it at the present point.

The construction of a *U*-calculus corresponding to a Q-like tense-logic also presents problems, but it would seem that it would contain *CTaNpNTap* but not its converse *CNTapTaNp*; that it would have both *CTaCpqCTapTaq* and *CTaKpqKTapTaq* and their converses; that *P*, *H*, *F*, and *G* would have to be dealt with separately by

TP: *ETaPpΣbKUbaTbp*
TH: *ETaHpΠbCUbaTbp*
TF: *ETaFpΣbKUabTbp*
TG: *ETaGpΠbCUabTbp*;

that the rule to infer $\vdash Ta\alpha$ from $\vdash\alpha$ would have to be replaced by one to infer $\vdash NTaN\alpha$ from $\vdash\alpha$, and perhaps also one (call it RTC) to infer $\vdash CTa\alpha Ta\beta$ from $\vdash C\alpha\beta$ if β had no free variables not in α; and that α would be a thesis in a tense-logic if and only if $NTaN\alpha$ were a thesis in the corresponding *U*-calculus. We could then derive, e.g. the rule corresponding to that to infer $\vdash NPN\alpha$ from $\vdash\alpha$, as follows:

1. $NTaN\alpha$
2. $NTbN\alpha$ (1 subst.; α is unaffected since, being a tense-logical formula, it contains no *a*'s)

[1] A. Mostowski, 'On the Rules of Proof in the Pure Functional Calculas of the First Order', *Journal of Symbolic Logic*, vol. 16, no. 2 (June 1951), pp. 107–11.

[2] pp. 45–47 and 46, cf. also p. 60.

3. $CUbaNTbN\alpha$ (2, $CpCqp$)
4. $\Pi bCUbaNTbN\alpha$ (3, UG)
5. $N\Sigma bKUbaTbN\alpha$ (4)
6. $NTaPN\alpha$ (5, TP)
7. $CTaNNpTap$ ($CNNpp$, RTC)
8. $NTaNNPN\alpha$ (6, 7, $CCpqCNqNp$).

Various special conditions on U could be imposed as before, but they would not have exactly the same consequences.

8. *Tensed predicate logic with now-empty names, in Cocchiarella, Rescher, and Hamblin.* Current work on tensed and modalized predicate calculi tends to avoid these problems by approaching the matter in another way, i.e. with a different 'rule of ampliation'. For instance, in the tensed predicate calculi of Cocchiarella it is boldly ruled that x, y, and z are the particular individuals they are even before and after they exist, and he has quantifiers that range over the whole bunch of them at all times. Identifiable individuals thus conceived *can* of course 'come into existence', and be brought into existence too, though it is questionable whether the latter would be seriously describable as creation out of *nothing*. It certainly doesn't fit Geach's formula for that, for when God gives existence, and human existence in particular, to one of these patients in the waiting room, we can say 'For some already given x, God brings it about that x is a man'. We can also answer Buridan's objections to $CP\Pi x\phi x\Pi xP\phi x$. For in the relevant sense of 'everything' it has never been the case (even on the Christian hypothesis) that everything is God—there always have been x's of which we could say 'Now *that* isn't God', though before the creation the only ones we could say this of would be still awaiting existence. The laws of this kind of tensed quantification are 'Barcanian' and there is just no question, in a system of this type, of revising the underlying *propositional* tense-logic at all.

If we have some means of symbolizing the form 'x now exists', we can define, in a system of this type, another sense of 'everything' namely 'everything that actually exists', in terms of which such objections as Buridan's could still be put up, but now the law to which he objects wouldn't have the form $CP\Pi x\phi x\Pi xP\phi x$ but rather $CP\Pi xC\psi x\phi x\Pi xC\psi xP\phi x$, 'If it has been that (everything that ψs ϕs) then everything that ψs has ϕd', and this isn't a law in *any* system; e.g. 'It has been that (everyone in room E is a gambler)' doesn't imply that everyone who is now in room E has been a gambler. For it could be that room E was once full of gamblers but now has some other people in it who have never gambled in their life. Similarly it could be that although the Existence room once had no one but God in it, now it has others in it too.

The use of '$E!x$' ('x actually exists') to define 'Everything-real' and 'Something-real' in terms of the unrestricted 'Everything' and 'Something' ('Everything-real ϕs' as 'Everything ϕs-if-it-exists' and 'Something-real ϕs' as 'Something exists-and-ϕs') is recommended by Rescher.[1] Cocchiarella reverses the procedure by adding an *undefined* restricted universal quantifier to his unrestricted one, defining the particular quantifiers in terms of the corresponding universals in the usual way, and then using the restricted particular quantifier to define 'x actually exists' as 'Something-real is identical with x'. However the restricted quantifiers are introduced, we can define the less restricted 'Something that exists or will exist will ϕ' as 'It will be that (something-real ϕs)' and similarly with the past; just as we would do with a Q-like system in which the non-existent is not allowed to be individually designated. Rescher is mistaken, however, in suggesting that his restricted quantifiers behave *exactly* as the Q-like ones do. They do, indeed, involve us in similar departures from Barcan-type principles for mixing quantification and tensing, but in pure quantification theory Rescher's and Cocchiarella's restricted quantifiers are much less well-behaved than the Q-type ones. What is done at this point, in fact, is to save standard tense-logic (and unrestricted detachment) by dropping standard quantification theory; for example, with wide-ranging names but restricted quantifiers it is no longer a law that if a ϕs then something ϕs, for maybe the only a that ϕs doesn't yet exist (and so doesn't count as 'something', in the sense of 'something real'). The *un*restricted quantifiers do of course have the standard laws.

Cocchiarella raises the question whether the unrestricted quantifiers are really needed, and decides that they are—rightly, it seems to me, given his comprehensive use of names. We have

[1] N. Rescher, 'On the Logic of Chronological Propositions', *Mind*, Jan. 1966.

seen that appropriately placed tense-operators and restricted quantifiers will suffice to define forms like 'Something that is or will be real will ϕ', and it might be thought that the unrestricted 'Something ϕs' could be defined quite generally as 'It either is or has been or will be that something-real ϕs'; and 'Everything ϕs' analogously. But if we permit x's for which 'x does not exist' is now true, there must surely be *some* sense of 'something' in which we can infer from this that something does not exist; but the proposed translation of this conclusion—'It is or has been or will be that (something-that-exists does not exist)'—is simply false.

Cocchiarella's system is consistent with using the form ϕa to cover not only assertions about what does not yet or does not any longer exist but also assertions about 'objects' which do not exist and never have existed and never will; though this interpretation could be precluded by introducing an axiom to the effect that 'everything' either exists or has existed or will exist. (Such an axiom would be easily formulable in Cocchiarella's system.) It was, indeed, suggested by Hamblin in 1958 that tense-logic needs three quantifiers—one corresponding to the liberal interpretation of Cocchiarella's 'possible' quantifier, one to its more restricted interpretation, and one to Cocchiarella's 'actual' quantifier; taking $\Sigma x\phi x$ in the first sense as primitive, he defined the third sense (in the manner of Rescher) as $\Sigma xKE!x\phi x$ and the second as $PF\Sigma xKE!x\phi x$. Other modifications are also possible; for example, Dana Scott has devised a system in which names can apply to things before and after as well as during their existence but before and after their existence individuals are indistinguishable (cf. Edwards).

In modal logic also, of course, we can avoid the complications of the system Q by quantifying over *possibilia*. In both areas, in fact, we have a choice between a certain amount of awkwardness and a certain amount of superstition. Presumably because the notion of a mere *possibile* is somewhat less 'tight' and logically demanding than that of a merely past or merely future individual, modal logicians have been more ready than tense logicians to accept solutions in which mere *possibilia* are included among the individuals that names may designate, but only the ϕ-ing of some (or all) *actual* individuals is allowed to verify the assertion that something (or everything) ϕs. Such a solution has been developed, for example, by Kripke, who minimizes the resulting mess in quantification theory by having no theses with free variables.[1] In particular, he does not and in his system cannot assert $\vdash C\Pi x\phi x\phi y$, 'If everything-real ϕs then y ϕs', which could yield a false proposition if the name of a mere *possibile* were put for y; he merely asserts $\vdash \Pi y C\Pi x\phi x\phi y$, i.e. 'It is true of anything real that if everything real ϕs then that thing ϕs'. This system is 'Myhillian' in the sense of having S5 and most of quantification theory but not $C\Pi xL\phi xL\Pi x\phi x$, but it achieves this only by a deliberate impoverishment of the formal machinery.

Kripke has, however, a suggestion in a footnote which could be developed in some interesting directions, both in modal logic and in tense-logic. The suggestion might be re-stated in the following way: Medieval logicians distinguished between predicates (like 'is red', 'is hard', etc.) which entail existence, and predicates (like 'is thought to be red', 'is thought of', etc.) which do not.[2] Suppose we use ϕ, ψ, etc., for predicates generally, and f, g, etc., for the former sub-class of predicates. f, g, etc., are substitutable for ϕ, ψ, etc., but not vice versa; and complexes like Nf, Mf, etc., are substitutable for ϕ, etc., but not for f, etc. (such complexes *are* predicates, but are not predicates entailing existence). Similarly with more than monadic predicates. What Kripke then says is that we could add to his axioms the 'closure' of the formula $CKfy\Pi x\phi x\phi y$, i.e. we could add $\Pi yCKfy\Pi x\phi x\phi y$. This, however, would be a redundant addition, since it follows in his system from $\Pi yC\Pi x\phi x\phi y$, which he already has. The more interesting thing that these new variables make possible would be the reformulation of his system *with* free variables, and with $C\Pi x\phi x\phi y$ replaced by the qualified form $CfyC\Pi x\phi x\phi y$. The restricted variables in fact offer another way of expressing the idea of existence—the last formula amounts to '*If y exists*, then if everything ϕs, y ϕs'. Given this axiom, the unqualified $C\Pi xfxfy$ and $Cfy\Sigma xfx$ are easily provable for the restricted predicates. We have

1. $CfyC\Pi x\phi x\phi y$

[1] S. A. Kripke, 'Semantical Considerations on Modal Logic', *Acta Philosophica Fennica*, Fasc. 16 (1963) pp. 83–94.

[2] See, e.g., W. Burleigh, *De Puritate Artis Logicae Tractatus Longior* (Franciscan Institute, 1955), pp. 57–58.

2. $CfyC\Pi xN\phi xN\phi y$ (1 $\phi/N\phi$)
3. $CfyC\phi yN\Pi xN\phi x$ (2, $CCpCqNrCpCrNq$)
4. $CfyC\phi y\Sigma x\phi x$ (3, Df. Σ)
5. $CfyCfy\Sigma xfx$ (4, ϕ/f)
6. $Cfy\Sigma xfx$ (5, $CCpCpqCpq$)
7. $C\Pi yfyC\Pi xfxfy$ (1 ϕ/f, $\Pi 1$)
8. $C\Pi xfxC\Pi xfxfy$ (8, re-lettering of bound variables)
9. $C\Pi xfxfy$ (8, $CCpCpqCpq$).

We could also use Cocchiarella's 'possible quantifiers' with the normal rules, and then use the restricted predicates to define '*x* exists' as Σffx, and so to define the 'actual' quantifiers in the usual way.

9. *Tensed ontology*. As was pointed out in *Time and Modality*, we can also keep a standard modal logic or tense-logic, and a very simple quantification theory too, if we just have no Russellian individual name-variables at all, bound *or* free, but only devices for referring to individuals obliquely, as in Leśniewski's 'ontology'. The awkwardness which *this* procedure forces upon us is a necessity for distinguishing operators which form complex predicates from ones which form the corresponding complex propositions. For example, where *a* and *b* do not stand for proper but for common names, 'For some *a* (it will be that (the *a* is a *b*))' is equivalent to 'It will be that (for some *a* (the *a* is a *b*))' (Barcan formula); but neither of these is equivalent to 'For some *a*, the *a* is a thing-that-will-be-a-*b*'. And more fundamentally, dropping the quantifier, 'It will be that (the *a* is a *b*)' is not equivalent to 'The *a* is a thing-that-will-be-a-*b*'. For the latter implies, but the former does not, that what will be a *b* now exists, since only what exists can properly be called 'The *a*'; or more accurately, the form 'The *a* is a *b*', whatever *b* might be (even if it is of the form 'thing-that-will-be-a-*b*') implies 'The *a* exists', i.e. 'The *a* is an object', or 'There is such a thing as the *a*'; but the form 'It will be that the *a* is a *b*' only implies 'It will be that there is such a thing as the *a*'. On the other hand, 'It will be that the *a* is a *b*' implies that what will be a *b* will be the *a* when it is a *b*, whereas 'The *a* is a thing-that-will-be-a-*b*' does not imply this (it may, for all that this tells us, have ceased to be the *a* by the time it is a *b*).

If we symbolize 'The *a* is a *b*' as ϵab, and the term 'object' as

V, we have as a law *CϵabϵaV*; and indeed the form *ϵaV*, 'The *a* is an object', can be *defined* as *Σbϵab* ('There is something that the *a* is'), so that our law amounts to the existential generalization *CϵabΣbϵab*. If we write *fb* for the term 'thing that will be a *b*', we obtain *CϵafbϵaV*, or *CϵafbΣbϵab*, by substitution for free *b*. But we do not have *CFϵabΣbϵab*, but only *CFϵabΣbFϵab* (and *CFϵabFΣbϵab*).

Again, 'It has not always been true that (the *a* exists)', *NHϵaV*, is equivalent to 'It has at some time been false that (the *a* exists)', *PNϵaV*; but the former is not equivalent to 'The *a* is a thing-that-has-not-always-existed', *ϵanhV*, nor the latter to 'The *a* is a thing-that-at-some-time-was-non-existent', *ϵapnV*, or *ϵapΛ*, writing *Λ* for *nV*, i.e. 'non-object', 'non-exister'; nor are these last two forms equivalent to one another. Taking the first point first: 'It has not always been that (the *a* exists)' does not say that any particular object has lasted for a finite time only, but rather that it is only for a finite time that anything at all has been 'the *a*'; whereas 'The *a* is a thing-that-has-not-always-existed' does say the first thing, but is compatible with 'The *a* exists' having always been true, though different things have been 'the *a*' at different times.

The other point is trickier. Note, firstly, that 'The *a* is a non-object', *ϵaΛ*, is always false, since anything of the form 'The *a* is a *b*' (even *ϵaΛ* itself) implies that the *a* is *not* a non-object, but an object (though 'It is not the case that the *a* is an object', 'There is no such thing as the *a*', *NϵaV*, which is *not* of the form *ϵab*, *is* sometimes true). Note, secondly, that it is a reasonable law that if the *a* is a thing-that-has-been-a-*b* then it has been the case that something is a *b* (though it may not then have been 'the *a*'), i.e. we have *CϵapbΣcPϵcb*, even if we don't have *CϵapbPϵab*. Hence, putting *Λ* for *b*, we have *CϵapΛΣcPϵcΛ*. But *ϵcΛ* has *always* been false, for any *c*, i.e. *NΣcPϵcΛ*, and so *NϵapΛ*, i.e. it cannot be the case that the *a* is a thing-that-has-been-a-non-object. On the other hand, it *can* be the case that the *a* is a thing-that-has-not-always-been-an-object, *ϵanhV*. It would seem that the tensing of terms is not only not definable by means of the tensing of propositions, but itself has something Q-like about it, however orthodox the tensing of propositions may be. Even if *PN* has the same force as *NH*, *pn* (as in *pnV*, i.e. *pΛ*) isn't interchangeable with *nh*.

It was also pointed out in *Time and Modality* that in tensed ontology there are some advantages in taking as undefined, not the 'weak' form ϵab, meaning 'The only thing that is *now* an *a* is now a *b*', but the 'strong' form $\epsilon' ab$, meaning 'An *a* which is the only thing *ever* to be an *a*, is a *b*'. Certain observations made there about these two forms were improved upon by Geach in 1957. A succession of increasingly shorter single axioms that were found by Leśniewski for untensed ontology included the following:

1. $\Pi a \Pi b E \epsilon ab K K \Sigma c \epsilon ca \Pi c \Pi d C K \epsilon ca \epsilon da \epsilon cd \Pi c C \epsilon ca \epsilon cb$ (1920)
2. $\Pi a \Pi b E \epsilon ab K \Sigma c K \epsilon ca \epsilon cb \Pi c \Pi d C K \epsilon ca \epsilon da \epsilon cd$ (1921)
3. $\Pi a \Pi b E \epsilon ab \Sigma c K \epsilon ac \epsilon cb$ (1929)

It was noted in *Time and Modality* that 1 does, but 3 does not, hold for the plain ϵ of tensed ontology; that 1 does not hold in tensed ontology if ϵ is replaced throughout by ϵ', though it does hold if this replacement is made only in the right-hand argument of the main equivalence, this giving us a way of defining ϵ in terms of ϵ'. Geach pointed out that what has just been said of 1 is equally true of the shorter formula 2 (from which 1 can be deduced, and vice versa, without appealing to any principles which hold in untensed but not in tensed ontology), and also that the shortest formula of all, 3, does hold if ϵ is replaced by ϵ'.

An ϵ-type function of tensed ontology which has still to be investigated, but which has some useful properties if taken as undefined, is the simple 'The only thing ever to be an *a* is now a *b*', this being understood *not* as implying that the thing is now an *a* (but still as implying that it either is or has been or will be one). Reinterpreting ϵ' in this way, the weak ϵ is still definable in terms of it, though not quite so simply as in terms of the ϵ' of *Time and Modality*. With the latter, we can equate

(α) The only thing that is now President of the United States is a Texan (ϵab)

with

(β) For some *c*, a *c* which is the only thing ever to be a *c* is now both President of the United States and a Texan ($\Sigma c K \epsilon' ca \epsilon' cb$), and

> For any *c* and *d*, if a *c* which is the only thing ever to be a *c*, and a *d* which is the only thing ever to be a *d*, is now President of the United States, then a *c* which is the only thing ever to be a *c* is now a *d* ($\Pi c \Pi d C K \epsilon' ca \epsilon' da \epsilon' cd$).

(At least, we can assert this equivalence if we assume that for every object there is some *c* which it now is and which nothing else has ever been.) If in this equivalence we drop the phrases 'a *c* which is' and 'a *d* which is', i.e. if we use the new ϵ', we cannot be sure of the 'is *now* a *d*' with which the equivalence finishes up; but the equivalence will hold if we replace it by 'is or has been or will be a *d*', i.e. if we make the last clause

$$\Pi c \Pi d C K \epsilon' ca \epsilon' da A A \epsilon' cd P \epsilon' cd F \epsilon' cd.$$

(This only assumes that for every object there is some *c* which it, but never anything else, is or has been or will be.) Or (using the stronger assumption) we could make the last clause

$$\Pi c \Pi d C K \epsilon' ca K \epsilon' cd \epsilon' dd \epsilon' cd.$$

(The $\epsilon' ab$ of *Time and Modality* is definable in terms of the new one as $K \epsilon' ab \epsilon' aa$.)

The definability of ϵ means that forms like ϵafb are definable in terms of forms like $\epsilon' afb$, and this is important because in these last the tensing of a term *can* be replaced by ordinary propositional tensing. We simply equate

> (α) The only thing ever to be an *a* is now a future-*b* ($\epsilon' afb$)

with

> (β) The only thing ever to be an *a* now exists ($\Sigma c \epsilon' ac$) and it will be that the only thing ever to be an *a* is a *b* ($F \epsilon' ab$).

The forms $\epsilon' anhV$, 'The only thing ever to be an *a* is a thing that has not always existed' and $\epsilon' apnV$, or $\epsilon' ap\Lambda$, 'The only thing ever to be an *a* is a thing-that-was-once-a-non-existent', can then be distinguished by equating the former with

> The only thing ever to be an *a* exists ($\Sigma c \epsilon' ac$, or $\epsilon' aV$) and it has not always been the case that the only thing ever to be an *a* exists ($NH \epsilon' aV$),

and the latter with

> The only thing ever to be an *a* exists, and it has been the case that the only thing ever to be an *a* is a non-existent ($P\epsilon' anV$),

this being in turn equivalent to

> The only thing ever to be an *a* exists, and it has been the case that (i) the only thing ever to be an *a* exists, and (ii) it is not the case that the only thing ever to be an *a* exists ($PK\epsilon' aVN\epsilon' aV$),

of which the last component is impossible. The point here, I think, is not so much that *nh* is different from *pn*, as that with complex terminal prefixes, juxtaposition is not associative; the forms we are really distinguishing are $\epsilon' a(pn)V$ and $\epsilon' ap(nV)$. The difference is roughly between 'thing that formerly did not exist' and 'thing that was formerly a non-existent'. It is an ambiguity like the scope-ambiguity which arises in Russell's theory of descriptions; and it would be of no importance if we could be sure that forms with complex terms are in this system entirely dispensable in favour of complex propositions. With complex predicate terms, this seems certainly so, and subject terms can always be put into the predicate position by means of the equivalence

$$E\epsilon' ab\Sigma cKK\epsilon' cbM\epsilon' ca\Pi dLC\epsilon' daM\epsilon' dc.$$

('The only thing ever to be an *a* is a *b* if and only if for some *c*, 1. the only thing ever to be a *c* is a *b*, 2. the only thing ever to be a *c* is or has been or will be an *a*, and 3. for any *d*, if ever the only thing ever to be a *d* is an *a*, then the only thing ever to be a *d* is or has been or will be a *c*.')

To give the meaning of the form $\epsilon' ab$ in Cocchiarella's system with free individual variables, we read *a* and *b* as verbs, and the whole becomes

> For some *x* now existing, it is now the case that *bx*, and it is or has been or will be the case that *ax*, and for any *y* that exists or ever has existed or ever will, it is and always has been and always will be the case that if *ay* then *Iyx*.

Note that both of Cocchiarella's kinds of quantifier are required here—an external 'particular actual' one and an internal 'universal possible'. The definition of the same form in a Q-type system presents difficulties which I do not at present know how to overcome. The last clause cannot be rendered

(A) It is and always has been (H) and always will be (G) that for any (real) y, if ay then Iyx,

for times of which it is not the case that x exists at them, will be times at which 'For any y, if ay then Iyx' is not true but unstatable, i.e. with transitory x's this clause (A) will never be satisfied. But neither can it be

(B) It is not and has never been (NP) and will never be (NF) that for some y, ay and $NIyx$,

for this can be satisfied too easily—it will be if the only times at which other things a'd were ones at which $NIyx$ was (through the absence of x) unstatable.

10. *Internal and external complexity in systems with free individual variables.* Having to distinguish between the formation of complex predicates and the formation of the corresponding complex propositions, is a complication which some writers have found it worth while to bear with even in systems which do contain free individual variables. One encounters it, in particular, in the development by G. E. Hughes and D. G. Londey of the logic of 'empty universes'.[1]

In beginning their treatment of first-order predicate logic, indeed, Hughes and Londey do without individual name-variables altogether, and simply form quantified propositions directly from predicates. Using their technique, but modifying their symbolism, we may write $\Pi\phi$ for 'Everything ϕs' and $\Sigma\phi$ for 'Something ϕs'. Complex predicates are formed in the same ways as complex propositions, so that we have forms like $\Sigma N\phi$ for 'Something doesn't-ϕ', to be contrasted with $N\Sigma\phi$ 'It is not the case that something-ϕs'; and $\Sigma K\phi\psi$, 'Something ϕ-s-and-ψs', to be contrasted with $K\Sigma\phi\Sigma\psi$, 'Something ϕs and something ψs'. In an empty universe $\Sigma\phi$ ('Something ϕs') is never true, while $\Pi\phi$ ($= N\Sigma N\phi$) always is, so that a logic which allows for such a possibility will lack the law $C\Pi\phi\Sigma\phi$.

[1] G. E. Hughes and D. G. Londey, *The Elements of Formal Logic* (Methuen, 1965), chs. 26 and 36.

When Hughes and Londey eventually introduce individual name-variables, they allow the form ϕx to have a meaning in an empty universe, or at least they allow the question as to whether it is or is not the case in such a universe that ϕx to be a genuine one, and they rule that in fact it is in such a universe *not* the case that ϕx, for any ϕ. This decision would seem to make the negation of ϕx, i.e. $N\phi x$, automatically *true* in such a universe; but if we allow $N\phi$ as a special case of ϕ, the same decision would seem to make $N\phi x$ automatically *false*. To avoid this contradiction, Hughes and Londey distinguish the form $N(\phi x)$, which *is* always true in an empty universe, from the form $(N\phi)x$, which is always false there. They have the law $C\phi x\Sigma\phi$, corresponding to $C\phi y\Sigma x\phi x$ in the usual systems, but not $C\Pi\phi\phi x$, corresponding to $C\Pi x\phi x\phi y$. The usual proof of the latter from the former by substitution and contraposition fails. Substitution takes us from $C\phi x\Sigma\phi$ to $C(N\phi)x\Sigma(N\phi)$, and then contraposition to $CN\Sigma(N\phi)N(N\phi)x$, but we cannot pass from this to $CN\Sigma N\phi\phi x$, as $N(N\phi)x$ does not imply $NN(\phi x)$, and so ϕx.

It is obvious that these devices could be used in tense-logic with individuals that exist at some times but not at others. In sketching such an extension I shall adopt a suggestion of Hughes—not, however, put forward in the book—for eliminating brackets. We simply write 'x ϕs', not as ϕx, but as $x\phi$, so that 'x is a non-ϕ-er' becomes $xN\phi$, while 'It is not the case that x ϕs' becomes $Nx\phi$. Similarly 'x is a thing-that-once-ϕd' can become $xP\phi$, while 'It was once the case that x ϕd' becomes $Px\phi$. 'x now exists', $xE!$, is definable as $xC\phi\phi$. $Cx\phi x\phi$ differs from this in being true of non-existents as well. In their predicate calculus for non-empty universes only, they have the axiom $CNx\phi xN\phi$, while in their predicate calculus for empty and non-empty universes alike, this is weakened to $C\Sigma\psi CNx\phi xN\phi$. In a logic to cope with *terms* which may be empty even when the universe is not, this needs to be further weakened to $Cx\psi CNx\phi xN\phi$ (the other would say in effect that x is either a ϕ-er or a non-ϕ-er, i.e. x exists, provided that *something* exists, even *something else*). From this (and $Cx\phi\Sigma\phi$) we can obtain $Cx\psi C\Pi\phi x\phi$, which may be compared with the formula $CfyC\Pi x\phi x\phi y$ in the Kripke-like logic mentioned at the end of Section 8.

In listing and in some cases proving some specimen laws of this sort of tense-logic, we may note that the system has a rule (call it Rx) that if α' and β' are constructed from predicate-variables in the same way as α and β are constructed from propositional ones, then if $\vdash C\alpha\beta$ in the propositional calculus, then $\vdash Cx\alpha'x\beta'$ in this predicate calculus. We now have

1. $Cx\phi xC\phi\phi$ — ($CpCpp$, Rx)
2. $Cx\phi xE!$ — (1, Df. $E!$)
3. $CxP\phi xE!$, — 'If x is a former-α-er, then x now exists' (2, $\phi/P\phi$)
4. $CPx\phi PxE!$ — 'If formerly x $\phi'd$, then formerly x existed' (2, RPC)
5. $CxP\phi Px\phi$, — 'If x is a former-ϕ-er, then formerly x ϕ'd' (see below)
6. $CxP\phi PxE!$, — 'If x is a former-ϕ-er, then formerly x existed' (5, 4, syll.).

Probably 5 needs to be laid down as a special axiom, though its analogue $CxN\phi Nx\phi$ is provable from the Hughes-Londey predicate-calculus basis as a theorem. Its converse $CPx\phi xP\phi$ ('If formerly x ϕ'd then x is a former-ϕ-er') is no more a law than is $CNx\phi xN\phi$ ('If x does not ϕ then x is a non-ϕ-er'); nor of course is $CPx\phi xE!$ ('If formerly x ϕ'd then x now exists); nor, though we have $C\Sigma P\phi P\Sigma\phi$ ('If something formerly-ϕ'd then formerly something-ϕ'd') do we have $CP\Sigma\phi\Sigma P\phi$ (Barcan formula: 'If formerly something-ϕ'd then something formerly-ϕ'd'). We have already seen that quantification theory in this system is a little eccentric; but it does seem to be another way of preserving standard propositional tense-logic. Its main defect is that there are difficulties in extending this type of symbolism beyond the monadic predicate calculus; but these may not prove insuperable.

11. *The difficulties of doing without non-existents.* One argument in favour of the view that if we are to use individual name-variables at all, we should let them cover non-existents, is that we often want to express *relations* between what now exists and what does not, e.g. that I am taller than my great-grandfather was. Comparisons of this sort, however, present problems even when they are not between objects that do not exist simultaneously. Take, e.g. 'I am fatter than I was', or its equivalent 'I used to be thinner than I am'. One thing that tense-logic is designed

precisely to facilitate is talk of persisting objects, and one thing that it is designed precisely to avoid is the introduction of pseudo-entities like 'me-at-t', 'me-at-t'', etc.; so a tense-logician will not want to make 'I used to be thinner than I am' express a comparison between such entities. But 'I used to be thinner than I am' certainly cannot mean 'It was the case that (I am thinner than I am)', since this is something that *never* was the case.[1] We have two choices here, it seems to me. If sizes and distances are absolute, we can say 'For some girths G and G', it was the case that my girth is G and it is the case that my girth is G', and G is (i.e. is-always) less than G'.' And if sizes and distances are relative, what we have is 'for some object x (e.g. the standard foot), it was the case that I am thinner than x, and it is now the case that I am not thinner than x.' The comparison between myself and my great-grandfather can, at least up to a point, be dealt with similarly—'It was the case that (for some x, x is my great-grandfather and is of height H), and my height is H', and H' is greater than H.'

There remains a difficulty about the compound 'It was the case that (for some x, x is my great-grandfather)', i.e. 'someone was my great-grandfather', or 'I am someone's great-grandson'. If we take the firm line with which we started, and admit no facts directly about non-existent individuals, and if y's great-grandfather ceased to exist before y started to, there cannot now be, or ever have been, any facts of the simple form 'x is y's great-grandfather'. However, we can analyse 'someone was y's great-grandfather' into a complex of relations between contemporaries in some such way as this: 'It was the case that (for some z, y is born to z, this resulting from the fact that it was the case that (for some w, w has intercourse with z, and it was the case that (for some u, z is born to u, this resulting from the fact that . . .)))', and so on; the whole being a fact directly about y only, and the aRb forms which enter into the component *general* facts (to the effect that it was the case that for some . . .) all expressing relations between contemporaries. But however we get around particular examples, it may well be

[1] I owe this simple puzzle to P. T. Geach. (It is related to one raised by Moore in his *Lectures on Philosophy*, p. 8, point (3).) The Schoolmen, it is worth noting, described relations as 'unreal', or as partly so, when either (*a*) they hold between objects which do not both exist when they hold or (*b*) they hold between an object and itself.

felt to be intolerable to have to deny that there are ever genuine ungeneralized relations between non-contemporaries.

12. *The admission of past existents but not future ones.* There is something to be said for a combination of solutions in which we are, broadly speaking, awkward about future objects and superstitious about past ones. Things that *have* existed do seem to be individually identifiable and discussable in a way in which things that don't yet exist are not (the dead are metaphysically less frightening than the unborn). Inhabitants of this half-way house can use names that refer to past and present objects only, and quantifiers that mean 'Something that is or has been' and 'everything that is or has been'. (Additional quantifiers restricted to what now is could be introduced to define 'exists' in Cocchiarella's manner, but that predicate could also be introduced in other ways, and the restricted quantifiers defined in terms of it.) This procedure would still eliminate, e.g. $CNPNCpqCPpPq$ but it would not assail its mirror image $CNFNCpqCFpFq$; e.g. if it will never be false that if nothing exists then this man doesn't exist, then if it will be that nothing exists it will be that this man doesn't; for now that he has come to be there will always be facts about him. Again, Buridan's objection to $CP\Pi x\phi x\Pi xP\phi x$ will stand, but one will not be able to make the same objection to $CF\Pi x\phi x\Pi xF\phi x$. 'If it will be that everything is God then everything will be God' will hold, if only because it cannot now *ever* be that *everything* is God—even after I cease to exist, I, for example, will be a countable exception to 'For all x, x is God'.

In sorting out the 'Barcan formulae' which would be true and false in this system, it is simplest to consider what happens when we combine the quantifiers with a specific Pn and Fn rather than with the generalized forms. We then have these laws for the past:

$$C\Sigma xPn\phi xPn\Sigma x\phi x$$
$$CPn\Sigma x\phi x\Sigma xPn\phi x$$
$$C\Pi xPn\phi xPn\Pi x\phi x$$

and these for the future:

$$C\Sigma xFn\phi xFn\Sigma x\phi x$$
$$CFn\Pi x\phi x\Pi xFn\phi x,$$

but we lack these three:

$$CPn\Pi x\phi x\Pi xPn\phi x$$
$$CFn\Sigma x\phi x\Sigma xFn\phi x$$
$$C\Pi xFn\phi xFn\Pi x\phi x.$$

(The reflection that the values of bound variables may receive additions but no deletions as time passes, makes it easy to work out these results intuitively.)

It has been suggested (e.g. by A. J. Kenny) that the naming of past individuals is easier than the naming of future ones merely because of the indeterminacy of the future. There can be facts directly about future individuals just as there can be facts directly about past ones, so long as their future existence is as definite as the past existence of the others is. I suspect, however, that this possible connexion between the subject of the present chapter and that of the last may be best exhibited in systems which do not use individual names at all but only the individualizing propositional forms of tensed ontology, embedded in something like a Peircean GHF logic rather than a standard GH one. Forms like $\epsilon'ab$, $P\epsilon'ab$, and $F\epsilon'ab$ would normally be taken as entailing that there is or has been or will be such a thing as the *a*, or 'the only thing ever to be an *a*'; we need perhaps a stronger ϵ, so used that the corresponding forms are all false unless there is or has been or *definitely* will be (strong Peircean *F*) such a thing as the *a* (e.g. '*XY*'s fourth child'). In such a logic, however, there will be complications not only when the only object that might satisfy such a description as 'the a' does not yet exist (as when it is not yet definite that *XY* will have a fourth child) but also when the description might be satisfied by some object that does exist, though it is not yet definite that it will be, or when it is not yet definite *which* presently-existing object will satisfy it. Some of these problems have already been discussed, in a preliminary way, in *Time and Modality*,[1] and there is nothing I could add now to what is said there. We know rather more today about indeterministic propositional tense-logic than was known in 1956, but not much more about tensed ontology.

13. *Summary of possible positions.* To sum up, this is still the untidiest and the most obscure part of tense-logic, though even

[1] pp. 101–3.

here the alternatives that are open to us are beginning to emerge with some clarity. We may (1) treat past and future alike, and given that we do so, we may (1.1) allow that there are facts directly about individuals of the form ϕa. Then we may (1.11) allow there to be such facts only about presently existing individuals, in which case our propositional tense-logic will be complicated in a Q-like way. If, however, we allow $\Sigma x\phi x$ to be a fact so long as there are facts of the form ϕa, our quantification theory will be of a standard sort (we will have both $\vdash C\phi y\Sigma x\phi x$ and $\vdash C\Pi x\phi x\phi y$); though detachment will have to be restricted. Or we may (1.12) allow that there are facts, of the form ϕa, which are directly about non-existent as well as existent individuals. This will give us a comparatively uncomplicated propositional tense-logic of one of the sorts discussed in Chapters III and IV. If (1.121) (Rescher and Cocchiarella) we still allow $\Sigma x\phi x$ to be a fact so long as there are facts of the form ϕa, i.e. if we allow non-existent as well as existent individuals to be values of bound variables, our quantification theory will again be of the standard sort. But $E!x$ will *not* be a law, and we shall need somehow to distinguish existent from non-existent individuals. We may (1.1211; Rescher) do this by an undefined function $E!x$, or we may (1.1212; Cocchiarella) introduce additional quantifiers such that $\Sigma x\phi x$ is only true if there are facts of the form ϕa in which a is existent; and the theory of *these* quantifiers will *not* be of the standard sort, but will lack either $C\phi y\Sigma x\phi x$ or $C\Pi x\phi x\phi y$ or both (normally both). Or we may (1.122), while allowing facts of the form ϕa directly about non-existent individuals, use *only* 'restricted' quantifiers of the sort just described (a procedure which is more of a live option in modal logic than in tense-logic). Or we may (1.2) *not* allow there to be facts *directly* about individuals, and use the a in ϕa for common names only, though one thing that this form could stand for might be 'There is exactly one a', and another might be 'The only a there is, is a b'. This will again give us a comparatively uncomplicated propositional tense-logic, and if we allow $\Sigma x\phi x$ to be a fact as long as there are facts of the form ϕa, a standard quantification theory also; but complex predicates (e.g. negative ones and tensed ones) may have to be formed in a different way from complex propositions. This last type of complication might also be accepted in alternatives 1.212 and

1.22 (Hughes and Londey). And finally, we might (2) treat past and future differently, with one type of solution for future-existers and a different one for past-existers; there are obviously many different ways of doing this.

I would like to finish, however, with a philosophical rather than a formal remark, though it may turn out to have a bearing on our formalisms. The problems of tensed predicate logic all arise from the fact that the things of which we make our predications, the 'values of our bound variables', include things that have not always existed and/or will not always do so. And this, I think, *is* a fact; it is unplausible to say either that the only things that are genuine individuals are 'ultimate simples' which exist throughout all time and merely get rearranged in various ways, or that there is only a single genuine individual (the Universe) which gets John-Smithish or Mary-Brownish in such-and-such regions for such-and-such periods. But the alternative to these two unsatisfactory theories has been presented in these pages a little too crudely; we are not really presented with a stark starting-to-be of an individual object with no antecedents whatsoever. Very roughly, countable 'things' are made or grow from bits of stuff, or from other countable 'things', that are already there. The precise logic of this process just hasn't been worked out yet, and until it has been, it seems likely that any tensed predicate logic can only be provisional in character.[1]

[1] For a rather unsatisfactory beginning of such an investigation, see A. N. Prior, 'Time, Existence and Identity', *Proc. Arist. Soc.*, 1965–6, pp. 183–92. (On p. 189, line 17, 'all times' should be 'that time'.)

APPENDIX A

POSTULATES FOR MODAL LOGIC, TENSE-LOGIC, AND *U*-CALCULI

(All postulates are for appending to propositional calculus with substitution and detachment.)

I. MODAL LOGIC

(Only systems of modal logic with established correlations with systems of tense-logic are included.)

§ 1. *Gödel–Feys systems* (*L* undefined)

§ 1.1. *Feys's (1950) system T* (= von Wright's M of § 2.1)

Df. M: $M = NLN$
RL: $\vdash \alpha \rightarrow \vdash L\alpha$
Axioms: 1. *CLCpqCLpLq*, 2. *CLpp*.

§ 1.2. *The system S4* (Gödel's axiomatization, 1933): T+*CLpLLp*.

§ 1.3. *The system S5* (Gödel, 1933): T+*CNLpLNLp* (or *CMLpLp*).

§ 2. *Von Wright's (1951) systems* (*M* undefined)

§ 2.1. *The system M* (= T of § 1.1):

Df. L: $L = NMN$
RL: $\vdash \alpha \rightarrow \vdash NMN\alpha$
RE: $\vdash E\alpha\beta \rightarrow \vdash EM\alpha M\beta$
Axioms: 1. *EMApqAMpMq*; 2. *CpMp*.

(If Ax. 1 is replaced by *CNMNCpqCMpMq*, RE may be dropped, and the equivalence to T made more obvious.)

§ 2.2. *The System M′* (= S4): M+*CMMpMp*.

§ 2.3. *The system M″* (= S5): M+*CMNMpNMp* (or *CMpLMp*).

§ 3. *Systems between T and S5*

§ 3.1. *The 'Brouwersche' system, or system B*: T+*CpLMp* (or *CMLpp*).

§ 3.2. *The system S4.2*: S4+*CMLpLMp* (simplified from Prior's *CMLpLMLp*; Geach 1957).

§ 3.3. *The system S4.3*: S4+*ALCLpqLCLqp* (simplified from Lemmon's *ALCLpLqLCLqLp*; Geach 1957), or *CKMpMqAMKpMqMKqMp* (Hintikka, 1957).

§ 3.4. *The 'Diodorean' system D*: S4.3+*CLCLCpLppCMLpp* (simplified from Dummett's *CLCLCpLpLpCMLpLp*, Geach 1959; completeness proved by Kripke 1963, and Bull 1963).

II. TENSE LOGIC

§ 4. *The minimal tense-logic* K_t (Lemmon, 1965)

§ 4.1. *With G and H undefined*:

Df. F: $F = NGN$ — Df. P: $P = NHN$

RG: $\vdash \alpha \rightarrow \vdash G\alpha$ — RH: $\vdash \alpha \rightarrow \vdash H\alpha$

Axioms:

1.1. *CGCpqCGpGq* — 1.2. *CHCpqCHpHq*

2.1. *CNHNGpp* — 2.2. *CNGNHpp*.

§ 4.2. *With F and P undefined*:

Df. G: $G = NFN$ — Df. H: $H = NPN$

RG: $\vdash \alpha \rightarrow \vdash NFN\alpha$ — RH: $\vdash \alpha \rightarrow \vdash NPN\alpha$

Axioms:

1.1. *CNFNCpqCFpFq* — 1.2. *CNPNCpqCPpPq*

2.1. *CPNFNpp* — 2.2. *CFNPNpp*.

Notes. (*a*) The 2's, in each case, are abbreviable to *CPGpp* and *CFHpp*, and could be replaced by *CpGPp* and *CpHFp*.

(*b*) With $L\alpha$ for $K\alpha G\alpha$, or $M\alpha$ for $A\alpha F\alpha$, the 'modal' fragment of K_t is the system T of § 1.1, or M of § 2.1.

(*c*) With $L\alpha$ for $KK\alpha G\alpha H\alpha$, or $M\alpha$ for $AA\alpha F\alpha P\alpha$, the 'modal' fragment of K_t is the system B of § 3.1.

§ 5. *Standard enlargements of the minimal system*

§ 5.1. *Axioms to be drawn upon, for addition to* K_t:

3. *CGpGGp* (= *CFFpFp* = *CHpHHp* = *CPPpPp* = *CFHpHp* = *CPpGPp* = *CPGpGp* = *CFpHFp*; Lemmon, 1965)

4. *CGGpGp* (= *GFpFFp* = *CHHpHp* = *CPpPPp* = *CHpHFp* = *CPGpPp* = *CGpGPp* = *CFHpFp*)

5.1 *CGpNGNp* (= *CNFNpFp* = *CNFpFNp* = *CGNpNGp* = *NGNCpp* = *FCpp*)

5.2. *CHpNHNp* (= *CNPNpPp* = *CNPpPNp* = *CHNpNGp* = *NHNCpp* = *PCpp*)

(Definitionally, 5.1. = *CGpFp* and 5.2 = *CHpPp*.)

6.1. *CKFpFqAAFKpqFKpFqFKqFp* (= *AGCpCGqrGCNpCGrq*; Lemmon, 1965) (= *AGCpCGpqGCGqp*; C. Howard, 1966)

6.2. *CKPpPqAAPKpqPKpPqPKqPp* (= *AHCpCHqrHCNpCHrq* = *AHCpCHpqHCHqp*)

7.1. *CKKpGpHpHGp* (= *CPFpAApFpPp*)

7.2. *CKKpHpGpGHp* (= *CFPpAApPpFp*).

§ 5.2. *System of 'The Syntax of Time-Distinctions'* (1954) *for dense, non-ending, non-beginning time*

Add 3, 4, 5.1, and 5.2 to K_t.

§ 5.3. *System for relativistic causal time* (Cocchiarella, 1965; superfluous axioms deleted).

Add 3 only to K_t.

With *L*α for *K*α*G*α, or *M*α for *A*α*G*α, the 'modal' fragment is S4 of § 1.2. or § 2.2.

§ 5.4. *System for linear time* (Cocchiarella, 1965; superfluous axioms deleted).

Add 3, 6.1 and 6.2 to K_t.

With *L*α for *K*α*G*α or *M*α for *A*α*G*α, the 'modal' fragment is S4.3 of § 3.3.

With *La* for *KK*α*G*α*H*α or *M*α for *AA*α*G*α*H*α, the 'modal' fragment is S5 of § 1.3 or § 2.3.

§ 5.5. *System for linear, non-ending, non-beginning time* (Scott, 1965).

Add 3, the 5's and the 7's to K_t.

'Modal' fragments as in § 5.4.

§ 5.6. *System for dense, linear, non-ending, non-beginning time* (Prior, 1965; superfluous axioms deleted).

Add 3, 4, the 5's and the 7's.

'Modal' fragments as in § 5.4.

§ 5.7. *System designed for the work of 5.6* (Hamblin, 1958) *and in fact giving logic of F as 'is or will be' and P as 'is or has been'*. (*F* and *P* undefined, dff. *G* and *H* as in § 4.2).

RG: $\vdash \alpha \rightarrow \vdash G\alpha$

RE: $\vdash E\alpha\beta \rightarrow \vdash EF\alpha F\beta$

RMI: In any thesis we may simultaneously replace every *F* by *P*, every *P* by *F*, every *G* by *H*, and every *H* by *G*.

Axioms: 1. *CGpFp*

2. *EFApqAFpFq*

3. *EFFpFp*

4. *EApPpGPp*

5. *EAApPpFpFPp*.

§ 5.8. *Equivalent system with G and H undefined.*

RG, RMI, and Axioms 1. *CGCpqCGpGq*, 2. *CGpp*, 3. *CGpGGp*, 4. *CpGPp*, 5. *CGpCHpGHp*; or simply add 2, 3, 5, and 5's image to K_t.

§ 6. *Systems for circular time*

§ 6.1. *Without change of sign*

Add to § 5.3 the axioms

1. *CGpHp* (= *CHpGp* = *CFpPp* = *CPpFp* = *CFGpp* = *CPHpp* = *CpGFp* = *CpHPp*; Lemmon, 1965).
2. *CGpp* (= *CHpp* = *CpFp* = *CpPp*).
3. *CGpGGp*.

§ 6.2. *With change of sign, and antipodes both past and future* (Lemmon 1965).

Add to K_t the axiom *CGpPp* (= *CHpFp* = *CGGpp* = *CHHpp* = *CpFFp* = *CpPPp*).

§ 6.2. *With change of sign, and antipodes neither past nor future.*

Add to K_t the axiom *CFGpPp* (Hamblin, 1965), which = *CPHpFp* = *CGpHPp* = *CHpGFp* = *CGGpHp* = *CHHpGp* = *CPpFFp* = *CFpPPp* = *CFGGpp* = *CPHHpp* = *CpGFFp* = *CpHPPp*.

§ 7. *Systems for the next moment* (*T*) *and the last moment* (*Υ*)

§ 7.1. *For appending to S5* (*of* § 1.3) *for undefined L* (*'always'*) (Scott, 1964):

1. *ELpTLp*
2. *ELpLTp*
3. *ETNpNTp*
4. *ETCpqCTpTq*
5. *ETΥpp*
6. *CLCpTpCLCqΥqCMKpqLApq*.

§ 7.2. *For use with G for 'It is and always will be' and H as 'It is and always has been'* (Lemmon, 1964).

Use RG and axioms 1 and 2 of § 5.8, and the axioms 3. *ETNpNTp*, 4. *ETCpqCTpTq*, 5. *ETGpGTp*, 6. *CGpTGp*, 7. *CGCpTpCpGp*, and 8. *ETΥpp*; and their mirror images.

§ 7.3. *For use with normal G and H* (Scott, 1965).

Add to the system of § 5.5 the axioms 1. *CGpTp*, 2. *ENTNpTp*, 3. *CTCpqCTpTq*, 4. *CpΥTp*, 5. *CTpCGCpTpGp*, and their mirror images.

§ 7.4. *For use alone* (Clifford, 1965):

RT: $\vdash\alpha \rightarrow \vdash T\alpha, \vdash \Upsilon\alpha$

Axioms: 1. *CTNpNTp*, 2. *CNTpTNp*, 3. *CTCpqCTpTq*, 4. *CpΥTp*, and their mirror images.

(Axioms with only *T* suffice for formulae with only *T*; ditto *Υ*.)

§ 7.5. *Equivalent of T-fragment with dyadic primitive* (von Wright, 1965).

(The primitive form is *Tpq* for '*p* now and *q* next'.)

RE: $\vdash E\alpha\beta \rightarrow \vdash Ef\alpha f\beta$ (for any *f* of the system).

Axioms: 1. *ETApqArsAAATprTpsTqrTqs*, 2. *EKTpqTrsTKprKqs*, 3. *EpTpAqNq*, 4. *NTpKqNq*.

III. *U*-CALCULI

§ 8. *Basic U–T postulates*

U1. *ETaNpNTap*

U2. *ETaCpqCTapTaq*

and for tense-logic

U3. *ETaGpΠbCUabTbp*

U4. *ETaHpΠbCUbaTbp*

and for modal logic

U5. *ETaLpΠbCUabTbp*.

(All for appending not merely to propositional calculus with substitution and detachment, but to first-order predicate calculus with identity.)

If and only if $\vdash\alpha$ in the modal system determined by RL: $\vdash a \rightarrow \vdash L\alpha$, and the axiom A1. *CLCpqCLpLq*, then $\vdash Ta\alpha$ in the system determined by U1, U2, and U5.

If and only if $\vdash\alpha$ in K_t of § 4, then $\vdash Ta\alpha$ in the system determined by U1, U2, U3, U4.

§ 9. *Correspondences between* U-*calculi, modal logics, and tense-logics* (mostly suggested by Lemmon)

We write '$\gamma \sim \beta$' for '$\vdash\alpha$ in the modal logic determined by RL, A1, and $\vdash\beta$, if and only if $\vdash Ta\alpha$ in the *U*-calculus determined by U1, U2, U5, and $\vdash\gamma$', or for '$\vdash\alpha$ in the tense-logic determined by $K_t + \vdash\beta$, if and only if $\vdash Ta\alpha$ in the *U*-calculus determined by U1, U2, U3, U4, and $\vdash\gamma$'.

§ 9.1. *Correspondence of formulae*

1. *Uaa* (reflexiveness)	~ *CLpp* (*CGpp*, *CHpp*)
2. *CUabUbb* (right reflexiveness)	~ *LCLpp* (*GCGpp*)
3. *CUabUba* (symmetry)	~ *CpLMp* (*CpGFp*)
4. *CUabCUbcUac* (transitiveness)	~ *CLpLLp* (*CGpGGp*)
5. *CUabCUacUbc*	~ *CMLpLp*
6. *CKUabUacAUbcUcb*	~ *CKMpMqAMKpMqMKqMp*
7. *CKUabUacΣdKUbdUcd* (convergence)	~ *CMLpLMp*
8. *CKUabUacAAIbcUbcUcb* (non-branching to right)	~ *CKFpFqAAFKpq-FKpFqFKqFp*

9. *CKUbaUcaAAIbcUbcUcb* (non-branching to left) ∼ *CKPpPqAPKpqPKpPqPKqPp*
10. *ΣbUab* (existence of successor) ∼ *CGpFp* (*CLpMp*)
11. *ΣbUba* (existence of predecessor) ∼ *CHpPp*
12. *ΣbKUabUba* ∼ *CGGpp* (*CLLpp*)
13. *CUabΣcKUacUcb* ('density') ∼ *CGGpGp* (*CLLpLp*).

§ 9.2. *Entailments of formulae*

(Where *n* and *m* are numbered conditions from § 9.1, '*n* entails *m*' means that (*a*) if the *U*-formula *n* is added as an axiom to the lower predicate calculus, *m* is deducible from it, and also that (*b*) if the modal formula *n* is added as an axiom to the modal system determined by RL and A1, or the tense-logical formula *n* is added to K_t, the modal or tense-logical formula *m* is deducible.)

1 entails 2, 10, 11, 12, 13.
2 entails 13, and is entailed by 1, 5, 6.
3 entails 7.
5 entails 2, 6, 8.
6 entails 2, 8, and is entailed by 5.
7 is entailed by 3.
10 is entailed by 1, 12.
11 is entailed by 1, 12.
12 entails 10, 11, and is entailed by 1.
13 is entailed by 1, 2.
(1+5) entails 3, 4.
(3+4) entails 5.
(1+6) entails 7.

§ 9.3. *Correspondence of systems*

('Condition *n*' means the *U*-condition *n* of § 9.1, supposed added to the basic postulates of § 8.)

System T of § 1.1	∼ condition 1
System S4 of § 1.2	∼ (1+4)
System S5 of § 1.3	∼ (1+5), i.e. (1+3+4)
System B of § 3.1	∼ (1+3)
System S4.2 of § 3.2	∼ (1+4+7)
System S4.3 of § 3.3	∼ (1+4+6)
System of § 5.3 (Cocchiarella, relativistic causal time)	∼ 4
System of § 5.4 (Cocchiarella, linear time)	∼ (4+8+9)

System of § 5.5 (Scott: doubly infinite linear time)	~ (4+8+9+10+11)
System of § 5.6 (Prior: dense Scott)	~ (4+8+9+10+11+13)
System of § 6.1 (normal circular time)	~ (1+3+4) (as S5)
System of § 6.2 ('east–west' circular time with antipodes)	~ 12.

APPENDIX B

MISCELLANEOUS FURTHER DEVELOPMENTS

1. *Von Wright's 'and next' and 'and then' calculi; 'and next' and metric tense-logic.* Von Wright's calculus for 'and next', sketched in Chapter IV, Section 3, has been supplemented by a calculus for another *Tpq* which he reads as '*p* and then *q*'. He means by this, not quite what Miss Anscombe means, but simply 'It is now the case that *p* and it will sooner or later be the case that *q*'. This does not seem a very idiomatic use of 'and then', except when the whole form is governed by some operator removing it from the actual present—'It was the case that (*p* now and *q* to come)' and 'It will be that (*p* now and *q* to come)' might well be read as 'It was the case that *p* and then *q*' and 'It will be that *p* and then *q*'; but we would hardly use this form for the simple '*p* now and *q* to come'. This, however, is not of much importance, and von Wright's function lends itself to fairly simple formal treatment.

Von Wright[1] repeats his axioms for 'and next', namely (in our modification of his symbolism)

A1. *ETApqArsAAATprTpsTqrTqs* A3. *EpTpAqNq*
A2. *EKTpqTrsTKprKqs* A4. *NTpKqNq*

(to be subjoined to propositional calculus with substitution, detachment and a rule of extensionality, licensing the interchange of logically equivalent expressions); and he notes that A2 may be replaced by the shorter *CKTpqTprTpKqr*. The postulates for 'and then' are the same, except that A2 has to be lengthened somewhat. Von Wright lists them, taken in this sense, as

B1. *ETApqArsAAATprTpsTqrTqs* B3. *EpTqAqNq*
B2. *EKTpqTrsTKprAAKqsTqsTsq* B4. *NTpKqNq*.

He notes that B2 may be replaced by the slightly shorter

B2′. *EKTpqTprTpAAKqrTqrTrq*,

or by the pair

B2.1. *CKTpqTprTpAAKqrTqrTrq*
B2.2. *CTpTqrTpr*.

[1] G. H. von Wright, 'And Then' (1966), the Comm. Phys. Math. of the Finnish Society of Sciences, vol. 32, no. 7 (1966).

It is clear that this new *Tpq* is easily definable within ordinary tense-logic as *KpFq*. Conversely, *Fp* is definable within the 'and then' calculus as *TCppp*, or as $T\tau p$, where τ is any arbitrary tautology. (A3 and A4, and of course B3 and B4, amount to $EpTp\tau$ and $NTpN\tau$ respectively.) Given these definitions, it is easy to show that the 'and then' calculus is equivalent to a future-tense calculus with the axioms (subjoined to propositional calculus with substitution, detachment, and the rule to infer $\vdash EF\alpha F\beta$ from $\vdash E\alpha\beta$):

F1. *EFApqAFpFq*
F2. *EKFpFqAAFKpqFKpFqFKqFp*
F3. *FApNp* (or $F\tau$)
F4. *NFKpNp* (or $NFN\tau$)

We get F1 from B1 by putting τ for *p* and *q* and dropping repeated disjuncts from the result; F2 and F4 from B2′ and B4 (in the form $NTpN\tau$) by putting τ for *p*; F3 from B3 by putting τ for *p* and detaching the first τ). Or we could replace F2 by the corresponding implication, plus *CFFpFp* (derivable as B2.2 $p/\tau, q/\tau$). The equivalence *ETpqKpTCqqq*, corresponding to the definition of *Tpq* in the *F* system as *KpFq*, is less summarily but still quite simply provable. And the converse derivations of the *T* postulates from the *F* ones are not difficult either. Von Wright himself equates his *T* system with a tense-logic (which he describes as a 'modal' logic) with *G* rather than *F* as primitive, and as postulates, 'duals' of the above (e.g. *EGKpqKGpGq* instead of F1). In both cases, the tense-logic is equivalent to the future-tense portion of Scott's for linear, transitive, infinite time, i.e. the system of § 5.5 of Appendix A.

For past and future together von Wright has a mirror image of the form *Tpq* which means '*p* now and *q* earlier', and which we may write here as *Ypq*. The full system has the *T* postulates and their images, and the pair of mixing axioms

$$ETpYqrAATKprqTpTrqKYprTpq$$
$$EYpTqrAAYKprqYpYrqKTprYpq.$$

If we equate *Pp* with $Y\tau p$, and *Ypq* with *KpPq*, and remember that $K\tau p = p$, the substitution p/τ in these gives

$$EFKqPrAAKrFqFKrFqKPrFq$$

and its image. These are the Cocchiarella axioms which Lemmon showed to be superfluous, together with their converses, from which we can derive the K_t theses *CpGPp* and *CpHFp* used in Lemmon's proofs. (The converse of *CFKqPrAAKrFq*, etc. entails that we have, among other things, *CKrFqFKqPr*, and therefore, by contraposition,

CNFKqPrNKrFq, and so *CGCqNPrCrNFq*, and so by substitution *CGCNPqNPqCqNFNPq*, and so *CqNFNPq* by detaching *GCNPqNPq*.) *CTKprqTpYqr* and its image, from which these K_t theses follow, could replace von Wright's longer pair.

For a two-way 'and next' calculus von Wright's 'mixing axioms' are *ETpYqrTKprq* and its image. These can, I think, be replaced by *CTpYqrTrq* and its image.

Both the 'and next' and the 'and then' calculus can be given axiomatizations that are more in the logical style of Feys's system *T* than of von Wright's *M*. The most compact postulate-sets of this sort which I have been able to find are those which follow; when von Wright's equivalences are expanded to pairs of implications it will be clear that the Feys-style postulates are more compact than his, though they do not lead so rapidly to normal forms and decision-procedures. I shall follow von Wright in putting an *A* before the 'and next' axioms and a *B* before the 'and then' ones. In both cases the axioms are to be subjoined to propositional calculus with substitution, detachment, and the rule

$$\vdash\alpha \rightarrow \vdash NTpN\alpha.$$

For 'and next' we have

A1. *CNTpNCqrCTpqTpr*

A2. *CpCTqrTpr* A4. *CpCNTpqTpNq* (or *TCppCpp*)

A3. *CTpqp* A5. *CTpNqNTpq*.

(The completeness of this basis is most simply shown by deducing from it Clifford's postulates for Scott's monadic *T*.) For 'and then' we have

B1. *CNTpNCqrCTpqTpr*

B2. *CpCTqrTpr* B4. *CpCNTpqTpNq* (or *TCppCpp*)

B3. *CTpqp* B5. *CTpTqrTpr*

B6. *CKTpqTprTpAAKqrTqrTrq*.

In 'reading off' these postulates, it may be noted that the form *NTpNq*, 'Not (*p* now and not-*q* next)' or 'Not (*p* now and not-*q* later)', is equivalent to 'If *p* now then *q* next' (first system) or 'If *p* now then *q* at all future times' (second system). A1 therefore amounts to 'If (if *p* now then if-*q*-then-*r* next) then if (*p* now and *q* next) then (*p* now and *r* next)', while B1 amounts to 'If (if *p* now then if-*q*-then-*r* at all later times) then if (*p* now and *q* later) then (*p* now and *r* later)'.

The A and B axioms have 1, 2, 3, and 4 in common; and B5 and 6 are simply von Wright's B2.2 and 2.1. Independence proofs for our A4, B4, A5, B5, and B6 are simple. A4 (= B4), *CpCNTpqTpNq*, is the only axiom which depends on time being supposed non-ending; it asserts in effect that if ever anything, say *p*, is true, then something—either *q* or not-*q*—is true later. The corresponding axiom in von Wright's set is *EpTpAqNq*. A5 is the only A-axiom which will not survive the reading of *T* as 'and then', and B5 the only B-axiom which will not survive the reading of *T* as 'and next'. A5, *CTpNqNTpq* (= *CTpqNTpNq*), assumes that there is some one interval by which *T* always takes us forward (if this were not so, *p*-now might be followed by *Nq* at one later time and by *q* at another); B5, *CTpTqrTpr*, assumes that there is not (if the interval mattered, it would matter that *TpTqr* only guarantees that we have *r two* steps later, while *Tpr* requires that we have it precisely *one* step later). The von Wright postulate with the same peculiarity as A5 is *EKTpqTrsTKprKqs*, or its abridgement *CKTpqTprTpKqr*. A5 (like its von Wright counterpart) also assumes that the future does not fork (if it did, we might have *Nq* one step along one fork and *q* one step along another); in the B set this is expressed, in the manner appropriate to unspecified intervals, by B6, with its obvious resemblance to the Hintikka-style linearity postulate for *F*.

B1, 2, and 3 and the *TY* thesis *CTKprqTpYqr*, with their mirror images, determine a *TY*-calculus which is equivalent, given the definitions of *F* in terms of *T* and vice versa, to the 'minimal' tense-logic K_t. The main relevant deductions in the *T*-calculus are

	1. *CNTτNCpqCTτpTτq*	(B1 *p*/τ, *q*/*p* *r*/*q*)
*	2. *CNFNCpqCFpFq*	(1, Df. *F*)
	3. *CTpqTτq*	(B2 *p*/τ, *q*/*p*, *r*/*q*; τ)
	4. *CTpqKpTτq*	(B3, 3, *CCpqCCprCpKqr*)
	5. *CKpTτqTpq*	(B2 *q*/τ, *r*/*q*, *CCpCqrCKpqr*)
	6. *ETpqKpTτq*	(4, 5)
*	7. *ETpqKpFq*	(6, Df. *F*).

(7 is the equivalence corresponding to Df. *T* in the *F* system.)

In the *A* system, while it is essential that *T* should take us forward by a single specific interval, this interval doesn't have to be an 'atom' of discrete time. As Rescher and Garson have observed,[1] the 'and next' system is interpretable within metric tense-logic with *Tpq* taken to mean that *p* is true now and *q* true after *any* specific interval, so long as the same interval is used throughout the system, i.e. *Tpq* = *KpFnq* for some constant *n*. The Feys-style postulates for this *T* are particularly easy to deduce from normal postulates for metric tense-logic, given this definition. A2 and A3, with *Tpq* expanded to *KpFnq*, become simple substitutions in the propositional-calculus laws

[1] Nicholas Rescher and James W. Garson, 'A Note on Chronological Logic', 1966, forthcoming in *Theoria*.

CpCKqrKpr and *CKpqp*; and A1, A4, and A5 use both the laws for *K* and laws for *Fn*. Our Feys-style B postulates (for 'and then') similarly follow very directly from standard sets for the 'topological' *F* (one doesn't have to go through the 'equivalential' variant of these given above).

Rescher and Garson also point out that, conversely, metric tense-logic may be developed within the 'and next' calculus, provided that only the integers are used in measuring intervals. There are various ways in which this may be done. If we use only non-negative integers, the future-tense portion of metric tense-logic may be developed within the 'and next' calculus by using the inductive definition

$$F0p = p$$
$$F(n+1)p = T\tau Fnp.$$

And once again, proofs are easier with the Feys-style axiomatization. For example, we prove FN2, *CNFnpFnNp*, as follows: For $n = 0$, this is just *CNpNp*. And given any *n* for which we have ⊢*CNFnpFnNp*, we may prove it for $n+1$. For $CNF(n+1)pF(n+1)Np$ expands to

$$CNT\tau FnpT\tau FnNp,$$

which we can prove syllogistically, since (*a*) we have $CNT\tau FnpT\tau NFnp$ by A4 p/τ, q/Fnp and detachment of τ, and (*b*) we have

$$CT\tau NFnpT\tau FnNp$$

from the inductive hypothesis by RT and A1. Similar, but simpler, inter-translations are possible with Scott's monadic *T*. ($Tp = Fnp$ and $F(n+1)p = TFnp$.)

A further point about von Wright's 'and next' system. When Kripke pointed out in 1958 that my *Time and Modality* matrix for Diodorean modality was not characteristic for S4, he also pointed out that a matrix which I gave at the same time for the system *M* or *T* was not characteristic for that system either. In this matrix, the elements are again sequences of 0's and 1's, the sequence for *Lp* having a 1 at a given place if and only if the sequence for *p* has a 1 both there and at the immediately succeeding place. (This verifies such non-*T* formulae as *CKMKpqMKpNqLp*.) In 1966 the same correction was made independently by K. Segerberg, who pointed out that an *Lp* for which this matrix *would* be characteristic would be one defined within von Wright's 'and next' system as *Tpp*, *Tpq* being conversely definable as *KpKMqCqLq* ($M = NLN$).

Segerberg's *L* is that of the Diodorean-modal fragment of a future-tense calculus in which *G* is equated with Scott's monadic *T*, and which therefore has the postulates RG, *CGCpqCGpGq*, *CNGpGNp*, and *CGNpNGp*. This corresponds to a *U*-calculus in which we have, beside U1, U2, and U3, the condition *CUabCUacIbc*. This condition entails but is not entailed by the non-branching condition

CUabCUacAAUbcUcbIbc,

and it neither entails nor is entailed by transitivity (cf. the resemblances and differences between von Wright's two *T*'s); it is satisfied, in a non-branching future, by the interpretation of *Uab* as '*a* is earlier than *b* by the specific interval *n*' (cf. Rescher's interpretation of von Wright).

2. *Minimal one-way tense-logic.* In Appendix A, § 8 and § 9, reference is made to a modal or quasi-modal system determined by RL: $\vdash\alpha \rightarrow \vdash L\alpha$ and A1. *CLCpqCLpLq*, subjoined to propositional calculus with substitution and detachment. This system has been called *T*(*C*) by E. J. Lemmon, to whom we owe the result that we have $\vdash\alpha$ in this system if and only if we have $\vdash Ta\alpha$ in the *U*-calculus determined by U1 (*ETaNpNTap*), U2 (*ETaCpqCTapTaq*), and U5 (*ETaLpΠbCUabTbp*), without any special conditions on *U* such as reflexiveness, transitivity, etc.[1] It is, in effect, the system *T* minus the axiom *CLpp*, and is equivalent to von Wright's system *M* minus the axiom *CpMp*. If we read *L* as *G*, it gives the purely future-tense fragment (and if we read *L* as *H*, the purely past-tense fragment) of the minimal tense-logic K_t.

3. *On the range of world-variables, and the interpretation of* U*-calculi in world-calculi.* In Chapter V, Section 6, it is suggested that we use variables *a*, *b*, *c*, etc., for instantaneous world-states, with the two axioms

A1. *Ma* A2. *ALCapLCaNp*,

where $M\alpha = AA\alpha P\alpha F\alpha$ and $L\alpha = KK\alpha H\alpha G\alpha$, and it is further suggested that, given the definitions *LCap* for *Tap* and *TbPa* for *Uab*, we should be able to prove the postulates of a given *U*-calculus from the corresponding tense-logic plus the above for 'worlds'. In particular, we should be able to prove the minimal *U*-postulates

U1. *ETaNpNTap* U3. *ETaGpΠbCUabTbp*
U2. *ETaCpqCTapTaq* U4. *ETaHpΠbCUbaTbp*

using the minimal tense-logic K_t. In that section, U1 and U2 are in fact thus proved, and the left-to-right implications in U3 and U4 are proved using a stronger tense-logic than K_t, in which we have such non-K_t theses as *CLpLHp*. It can now be shown that the unsolved problems here (proving U3 and U4 using only K_t and A1 and A2) are not soluble as stated, but are soluble in slightly modified forms.

[1] E. J. Lemmon, 'Algebraic Semantics for Modal Logics I', *Journal of Symbolic Logic*, vol. 31, no. 1 (March 1966), pp. 46–65.

It will be useful to begin with an objection which might be made to the postulate A1.[1] We may observe that A1 is equivalent to *CNaAPaFa*, 'If it is not now the case that *a*, then it either has been or will be the case that *a*', and there is something a little curious about this. If *a* is a world-state-proposition it will be a very 'strong' proposition in the sense of implying a great deal (as, indeed, A2 makes clear), so that its negation will be a very weak one, implying very little; and yet, by A1 as rephrased, it is strong enough to imply *APaFa*, and this seems a very substantial consequence. Again, are there not many possible combinations of the world's elements that may very well never be realized, and ought not our *a*'s, *b*'s, etc., to stand for these too?

There is much that needs to be disentangled here; though we shall see the objector has a point. In the first place, we should avoid the temptation to think of world-propositions as being singled out from others in virtue of their *form*, or as having a certain extensiveness of intuitive content (as asserting that so-and-so, the 'so-and-so' being a conjunction whose conjuncts are or could be all facts about what is, what has been, and what will be). This conception of a world-proposition (I start with it myself) has some usefulness, but we must get away from it in the end. In the second place, we must avoid confusing the rather artificial sense which we here assign to *M* with 'possibility' in some ordinary modal sense, e.g. logical possibility. These misconceptions are connected; it is only propositions with this vast content which will be both (*a*) 'possible' in the ordinary modal sense, and (*b*) '*L*-complete' in the modal sense. And if we were considering a calculus involving *both* ordinary modal *and* tense-logical notions, it would certainly be necessary to divide possible total world-states into ones which are realized at some time or other and ones which are never realized. But in Chapter V, Section 6, and here, we are considering a calculus which has no provision for the expression of ordinary 'logical' possibility, so that if we are to consider world-states other than the actual one, they must be ones whose relation to the actual one is expressible in tense-logical terms, and here *AAaPaFa* gives the *a*'s almost (though as we shall see not quite) as broad as range as we can get. Also, in a purely tense-logical calculus we cannot measure the relative 'strength' and 'weakness' of propositions by their content or by what they *necessarily* imply; we can only say that *p* is 'stronger' than *q* if *p* is *at no time* the case without *q* being the case, although *q* is *at some time* the case without *p* being the case; so that even a proposition with very little content could still be 'strong' in this sense if it happened to be very seldom true.

A world-state proposition in the tense-logical sense is simply an

[1] It was in fact made to me by Mr. Richard Campbell, of Magdalen College.

index of an instant; indeed, I would like to say that it *is* an instant, in the only sense in which 'instants' are not highly fictitious entities. To be the case *at* such-and-such an instant is simply to be the case *in* such-and-such a world; and that in turn is simply to be the case *when* such-and-such a world-proposition is the case. In this sense of 'instant' it is a tautology that a world-proposition is true at one instant only (it is true only when *that* world-proposition is true) and so is as 'strong' as any proposition that is ever true can be; though if time is circular it will not follow from this that if a world-proposition *is* true it neither has been nor will be true (for in circular time, 'has been' and 'will be' bring us back eventually to *the same* instant). It is also a tautology that anything that is true at one instant only will serve as a world-proposition; for however trivial its content it will be 'strong' enough to imply permanently whatever is true, i.e. it will never be true without all the things that are true-at-that-instant being true (either it will be false, or those things will be true along with it—their being 'true at that instant' just *is* their being true along with this proposition).

(I ought to remark here that my desire to sweep 'instants' under the metaphysical table is not prompted by any worries about their punctual or dimensionless character but purely by their abstractness. That some things are 'instantaneously true' I do not doubt,[1] and 'p now instantaneously' is an assertion easily expressible in Kamp's calculus of Φ and Ψ. It amounts to 'p now but Np just before and just after now', i.e. $KKpH'NpG'Np$, where $H'p = \Phi Cppp$, 'p throughout the interval between some past time and now'; and $G'p = \Psi Cppp$. This use of H' and G' was pointed out to me in 1965 by Richard Harschman, before their definition by Kamp in terms of his functors. But 'instants' as literal objects, or as cross-sections of a literal object, go along with the picture of 'time' as a literal object, a sort of snake which either eats its tail or doesn't, either has ends or doesn't, either is made of separate segments or isn't; and this picture I think we must drop. Cf. Chapter IV, Section 7.)

If we are to use the above conception of 'worlds' and 'instants' to identify the values of the variables for earlier and later instants in a U-calculus with the values of the variables for 'worlds' in a calculus that has them, our axioms for worlds ought to give us exactly one world for each element in the domain of the relation U. In fact the axioms A1 and A2 of Chapter V, Section 6, do not quite do this. A1, Ma, secures that each world-proposition is true at some time or other; but there ought also to have been a postulate securing that

[1] See the argument in Broad's *Examination of McTaggart's Philosophy*, vol. ii, pp. 273–5, substantially reproduced in Miss Anscombe's 'Before and After' (*Philosophical Review*, Jan. 1964), sect. 8 (pp. 17 ff.).

every point of time has a world-state which 'occupies' it (or which it is). What we want here is simply Σaa, from which we derive $L\Sigma aa$ by RL. We may call Σaa A3, but in what follows we shall only be using a derivative of it, asserting that if anything is ever true there is some world-state 'in' which it is true, i.e. which permanently implies it: $CMp\Sigma aLCap$. We may call this A3′. In deriving it from A3 we use the Barcan formula (BF) for worlds, which we know we have when given at least the system B for L and M. (We are concerned here with the gearing of tense-logics to U-calculi of the standard sort.) We also make use of the theorem of quantification theory that although we do not have $CK\Sigma x\phi x\Sigma x\psi x\Sigma xK\phi x\psi x$ ('If both something ϕs and something ψs then something both ϕs and ψs'), we do have $CK\Sigma x\phi xp\Sigma xK\phi xp$ ('If something ϕs, and p is the case, then there is something such that: it ϕs, and p is the case'). The proof is:

C	(1)	Mp	
K	(2)	$L\Sigma aa$	(A3, RL)
K	(3)	$MK\Sigma aap$	(1, 2, $CKMpLqMKpq$)
K	(4)	$M\Sigma aKap$	(3)
K	(5)	$\Sigma aMKap$	(4, BF)
K	(6)	$\Sigma aNLCaNp$	(5)
	(7)	$\Sigma aLCap$	(6, A2).

We cannot expect to derive a U-calculus within the corresponding tense-logic plus the calculus of worlds, unless the latter has A3 added to it. (A3 is needed, we shall find, for the derivation of the right-to-left implications in U3 and U4.) Even with this addition, moreover, we cannot expect to do it in any tense-logic but one for *linear* (non-forking) time, if we define $M\alpha$ as $AA\alpha P\alpha F\alpha$ and $L\alpha$ as $KK\alpha H\alpha G\alpha$. I have found it simplest to make this point clear to myself by associating U-calculi with linear diagrams. In the world-calculus within which (together with some tense-logic) we wish to develop the U-calculus, we want Lp to mean in effect that p is true *all over the diagram*, and Mp that p is true *somewhere* in the diagram. And it is only in linear time that p is true-somewhere-on-the-diagram if and only if it either is true now, or has been true, or will be true. For consider a non-linear diagram such as the following, where Fp is true at a given point if and only if p is true at some connected point towards the right, and Pp if and only if p is true at some connected point towards the left:

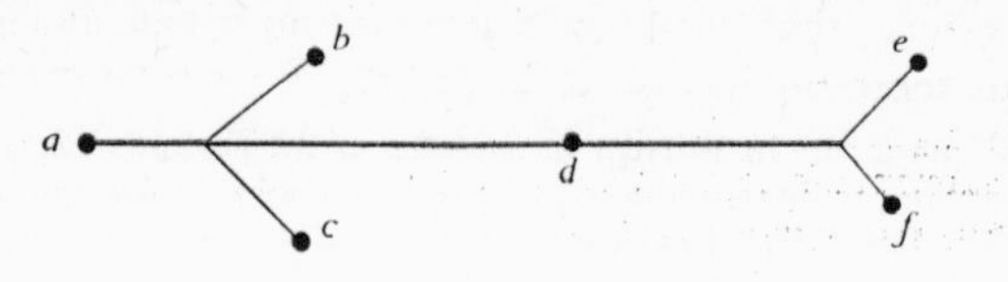

Let the world *d* be what is the case now. Then each of *a*, *d*, *e*, and *f* satisfy the condition 'It either is true now (like *d*) or has been true (like *a*) or in some future will be true (like *e* and *f*)'. But the 'could-have-beens' *b* and *c* do *not* satisfy this condition, although they *are* 'true-somewhere-on-the-diagram'.

Further, we can only be sure that even *a*, *e*, and *f* satisfy the condition if the earlier-later relation is taken to be transitive. Transitiveness seems, indeed, to be assumed in any representation of futurity by being 'somewhere' to the right and of pastness by being 'somewhere' to the left. The only non-transitive models of the earlier-later relation with which I am acquainted are ones in which this element of pure *direction* is supplemented by one of *distance*. In Hamblin's circular model, for instance, *FFp* ceases to give *Fp* if we go *too far* around the circle. *FFp* also takes us too far to give *Fp* if we read the latter as '*p* is *just* about to happen', i.e. as Scott's *Tp* or von Wright's '*Cpp* and *next p*'. In any case it is clear that our worlds will fail to 'cover the diagram' unless we do have *CFFpFp* (and consequently CF^npFp for each *n*), if all we say to give them a place on the diagram is *AAaPaFa*.

Is there any alternative but still tense-logical definition of *M* which *will* be satisfied by whatever is true at any point on a time-diagram, even in the absence of special assumptions about the character of the earlier-later relation? If so, this is obviously the *M* which should be used (with the corresponding *NMN* as *L*) in the postulates for 'worlds'. As a step towards such a revision, it may be pointed out that we certainly do not need to postulate complete linearity in order to find *a* tense-logical function which will be satisfied by whatever is true anywhere on the diagram. For even if we do have forking, provided that we have it in one direction only, and also have transitivity, anything true at any point on the diagram will satisfy a function *M′* which, if we still used $M\alpha$ for $AA\alpha P\alpha F\alpha$, would be defined as *MM*. We then have

$$\begin{aligned} M'p &= AA(AApPpFp)(PAApPpFp)(FAApPpFp) \\ &= AA(AApPpFp)(AAPpPPpPFp)(AAFpFPpFFp) \\ &= AAApPpFpPFp \end{aligned}$$

Here the second line comes from the first by *EPApqAPpPq* and *EFApqAFpFq*, which are in K_t, and the third from the second by (*a*) re-grouping disjuncts and dropping repeats, and (*b*) dropping disjuncts which imply ones that are already present (*CCpqEAApqrAqr*), either by *CPPpPp* and *CFFpFp* (transitivity) or *CFPpAApPpFp* (non-forking in the past—see Chapter III, Section 7). Similarly the new $L'p = KKKpHpGpHGp$. In the diagram given above, the new *M′p* covers the 'could-have-been' worlds *b* and *c*, for which we have *PF* (it was the case that they would be found in what was then one of the 'futures'). And any possible sequence of *P*s and *F*s by which we can be taken around the diagram will imply one or other of the four disjuncts in our new M', e.g. we have $CPFPpM'p$ by

$$\begin{aligned} PFPp &\rightarrow P(AApPpFp), \text{ by } CFPpAApPpFp \text{ and RPC;} \\ &\rightarrow AAPpPPpPFp, \text{ by } EPApqAPpPq; \\ &\rightarrow APpPFp, \text{ by } CPPpPp \\ &\rightarrow AAApPpFpPFp, \text{ by } CpAqp, \text{ etc.;} \end{aligned}$$

and $CFPFpM'p$ by

$$\begin{aligned} FPFp &\rightarrow AAFpPFpFFp \\ &\rightarrow AFpPFp \rightarrow M'p. \end{aligned}$$

Forking in *both* directions, however, as in the 'relativistic causal' time-sequence discussed by Cocchiarella, could take us to 'times' not covered by this M'. For this could give us alternative pasts as well as alternative futures, and such patterns as

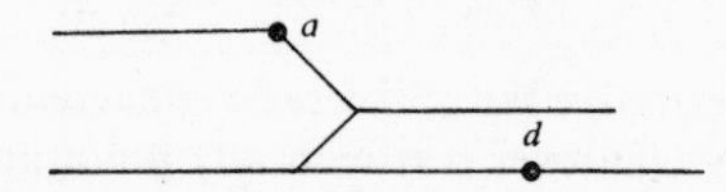

which might represent a $PFPa$ where we do not have either a (a is not true now), or Pa (a has not been true), or Fa (a will not be true), or even PFa (it has not been that a in what was once a future). One way of covering cases of this sort, and harder ones, is by introducing into our formal calculus the numerical superscripts that are often informally used by modal logicians for repeated *M*'s and *L*'s, as in $CLpL^2p$ for $CLpLLp$, and we could define a new $M''\alpha$ as $\Sigma nM^n\alpha$, where $M^1\alpha = AA\alpha P\alpha F\alpha$, and $M^{n+1}\alpha = M^1M^n\alpha$. This works in non-transitive worlds too. Even in K_t we have the metatheorem that if ϕ is some sequence of *P*s and *F*s, there is some n such that ϕp logically implies $M^n p$. For example, $CFPFpMMMp$ is provable even in K_t, since

$$\begin{aligned} MMMp &= A \ldots\ldots F(A \ldots\ldots PApPpFp) \\ &= A \ldots\ldots A(F \ldots\ldots FPApPpFp) \\ &= A \ldots\ldots A(A \ldots\ldots AFPpFPPpFPFp), \end{aligned}$$

and the same repeated use of $EFApqAFpFq$ and $EPApqAPpPq$ will secure the required result with longer sequences of *M*s. In general if ϕ has n symbols in it, ϕp implies $M^n p$ (and so $\Sigma nM^n p$). Similarly, if ψ is any sequence of *H*s and *G*s, and has n symbols in it, $L^n p$ (and so $\Pi nL^n p$) implies ψp. Intuitively, M^n is 'It is or has been or will be that it is or has been or will be that . . .' till n repetitions are reached;

while $\Sigma nM^n\alpha$ is 'It is or has been or will be that α, or else it is or has been or will be *that* it is or has been or will be that α, or else it is or has been or will be that (the preceding) . . . ' and so on *ad infinitum.*

These considerations suggest two lines of advance in this area. In the first place, we may retain our original simple definition of $M\alpha$ as $AA\alpha P\alpha F\alpha$, but only expect to get results with this if a transitive and linear tense-logic is used. This being understood, we can be satisfied with the proofs already found for the left-to-right implications in U3 and U4, and can start trying to prove the right-to-left ones; and also, of course, the condition of linearity on *U* (transitivity we have in fact done). We can also attempt to prove U1–U4, and the appropriate conditions on *U*, in particular weaker tense-logics with the definitions of *M* suitably adjusted; e.g. in transitive partly forking time using *M'*. I shall not attempt any of these proofs at this point, but will do something a little like it, but simpler. There is an even more *simpliste* definition of $M\alpha$ than $AA\alpha P\alpha F\alpha$ which only 'covers the diagram' if we take time to be not merely linear but circular, in the most straightforward sense (i.e. without change of sign at the antipodes). For this we can have $M\alpha = P\alpha = F\alpha$, and our underlying tense-logic will be simply S5 for this *M*. The proofs given in Chapter V, Section 6, of course still go through, including the proof of *CUabCUbcUac* (transitivity). This leaves us with the right-to-left implications in U3 and U4, and *CUabUba* (symmetry) and *Uaa* (reflexiveness) still to be proved. Our definition of *Uab* as *TbPa* is now equivalent to *TbMa*, i.e. *LCbMa*, so *Uaa* is *LCaMa*, which we obtain in S5 from *CpMp* and RL. *CUabUba*, i.e. *CLCbMaLCaMb*, we may prove *ad absurdum* thus:

C (1)	*LCbMa*	
C (2)	*NLCaMb*	
K (3)	*LCaNMb*	(2, A2)
K (4)	*LCMaMNMb*	(3, *CLCpqLCMpMq*)
K (5)	*LCMaNMb*	(4, *CMNMpNMp*)
K (6)	*LCMaNb*	(5, *LCNMpNp*)
K (7)	*LCbNb*	(1, 6)
(8)	*NMb*	(7, *CLCpNpNMp*),

which contradicts A1 (so that the combination of (1) and (2) is unallowable). Given this result, U4 follows from U3 (since it only differs from U3 in having *Uba* where the latter has *Uab*), and we prove the right-to-left implication *CΠbCUabTbpTaLp* by first proving the lemma *CTMapTaLp* thus:

C (1)	*LCMap*	
K (2)	*LCLMaLp*	(1, *CLCpqLCLpLq*)

(3) *LCaLp* (2, *LCpLMp*)

and then proving *CΠbCUabTbpTMap* by *reductio ad absurdum* as follows:

C (1) *ΠbCLCbMaLCbp*
C (2) *NLCMap*
K (3) *MKMaNp* (2, *ENLCpqMKpNq*)
K (4) *ΣbLCbKMaNp* (3, A3′)
K (5) *ΣbLCbKNpp* (4, 1)
(6) *ΣbNMb* (5),

which contradicts *ΠbMb* (from A1 and RL).

The second line of advance is to prove U1–U4 using K_t with a radically modified *M*, perhaps with the *M″* mentioned above. The logic of this function, and of the corresponding *L″*, is S5, even within K_t. For it can be shown that the laws of *L″*, i.e. ΠnL^n, include the analogues of RL, *CLCpqCLpLq*, *CLpp*, *CLpLLp*, and *CMLpp* (the last two together giving *CMLpLp*: $ML \rightarrow MLL \rightarrow L$). The infinity of the range of *n* gives us $C\Pi nL^n p\Pi nL^n\Pi nL^n p$ (though it doesn't, we may notice, give us the plain $CL^n pL^n L^n p$; we cannot get this without $CL^1 pL^1 L^1 p$, which is not provable in K_t). We have all of the others for $L^1\alpha$, i.e. $KK\alpha H\alpha G\alpha$, and it can be shown that if we have $\vdash\alpha \rightarrow \vdash L^n\alpha$, $CL^n CpqCL^n pL^n q$ and $CL^n pp$ for any *n* we have them for $n+1$ also. We get $\vdash\alpha \rightarrow \vdash LL^n\alpha$ from the hypothesis $\vdash\alpha \rightarrow \vdash L^n\alpha$ together with $\vdash\alpha \rightarrow \vdash L\alpha$ (with $L^n\alpha$ for our α), and with the axioms we have

1. $CL^n CpqCL^n pL^n q$ (Hyp.)
2. $CL^n pp$ (Hyp.)
3. $CLL^n CpqLCL^n pL^n q$ (1, RLC)
4. $CLCL^n pL^n qCLL^n pLL^n q$ (*CLCpqCLpLq*, subst.)
* 5. $CLL^n CpqCLL^n pLL^n q$ (3, 4, Syll)
6. $CL^n LpLp$ (2, *p*/*Lp*)
* 7. $CL^n Lpp$ (6, *CLpp*, Syll).

From these results (and quantification theory) it is clear that if α is a law so is $\Pi nL^n\alpha$ (at least, if the system has the symbolism), that $\Pi nL^n Cpq$ implies that $\Pi nL^n p$ implies $\Pi nL^n q$, and that $\Pi nL^n p$ implies *p*. $C\Sigma nM^n\Pi nL^n pp$ is slightly more complicated. We have, first, the following inductive proof of $CM^n L^n pp$, given that we have *CMLpp* for M^1 and L^1:

1. $CM^n L^n pp$ (Hyp.)
2. $CM^n L^n LpLp$ (1, *p*/*Lp*)
3. $CMM^n L^n LpMLp$ (2, RMC)
4. $CMM^n L^n Lpp$ (4, *CMLpp*, Syll).

Now $C\Sigma nM^n\Pi nL^npp$ is equivalent to $C\Sigma mM^m\Pi nL^npp$, which in turn is equivalent to $\Pi mCM^m\Pi nL^npp$, which we prove thus:

ΠmC (1) $M^m\Pi nL^np$
K (2) M^mL^mp (1, UI)
(3) p (2, lemma just proved).

One by-product of this result is the removal of a possible charge of arbitrariness against our proposed definition of the form *Uab*, '*a* is an earlier world than *b*'. Why *TbPa*, 'It is true at *b* that the world-state was formerly *a*', rather than *TaFb*, 'It is true at *a* that the world-state will eventually be *b*'? Given that the *L* in terms of which *T* is defined is one for which we have S5, and that it implies all sequences of *G* and *H*, we can prove the equivalence of these two forms; one implication in it by *reductio ad absurdum* thus:

C (1) *LCbPa*
C (2) *NLCaFb*
K (3) *LCaNFb* (2, A2)
K (4) *LCaGNb* (3, *NF* = *GN*)
K (5) *LLCaGNb* (4, *CLpLLp*)
K (6) *LHCaGNb* (5, *CLpHp*, RLC)
K (7) *LCPaPGNb* (6, *CHCpqCPpPq*, RLC)
K (8) *LCPaNb* (7, *CPGpp*)
K (9) *LCbNb* (8, 1)
(10) *NMb* (9),

which contradicts A1. (The converse implication is proved similarly.) A consequence of this equivalence is that our proof in Chapter V, Section 6, of $CTaGp\Pi bCUabTbp$, i.e. $CLCaGp\Pi bCLCbPaLCbp$, may, be paralleled by an exactly similar proof of $CTaHp\Pi bCUbaTbp$, which we may now equate with *CLCaHpCLCbFaLCbp*. Both proofs, moreover, involve only laws which we do have for the *L* we are now employing (in particular, we have *CLpLHp* and *CLpLGp*, from *CLpLLp*, *CLpHp*, and *CLpGp*).

Another equivalence which we may prove with our new *L* is that of *TaGp* with *TPap*. That the former implies the latter is provable as follows:

C (1) *LCaGp*
K (2) *LHCaGp* (1, *CLpLHp*)
K (3) *LCPaPGp* (2, *CHCpqCPpPq*, RLC)
(4) *LCPap* (3, *CPGpp*)

and that the latter implies the former, as follows:

C (1) *LCPap*

K (2) *LGCPap* (1, *CLpLGp*)
K (3) *LCGPaGp* (2, *CGCpqCGpGq*, RLC)
(4) *LCaGp* (3, *CpGPp*).

Given this equivalence, we may equate the right-to-left implication in U3, i.e. the implication of *TaGp* by *ΠbCUabTbp*, with *CΠbCUabTbpTPap*, which we prove *ad absurdum* thus:

C (1) *ΠbCLCbPaLCbp*
C (2) *NLCPap*
K (3) *MKPaNp* (2)
K (4) *ΣbLCbKPaNp* (3, A3)
K (5) *ΣbLCbKNpp* (4, 1)
(6) *ΣbNMb* (5),

contradicting *ΠbMb* (cf. the analogous proof of *CΠbCUabTbpTaLp* in circular time). The right-to-left implication in U4 may be proved similarly. Since our original proofs of U1 and U2 assume nothing for *L* that is not in Feys's *T*, and so can be carried through with this *L* also, we have now shown how to develop the entire minimal *U*-calculus within the world-calculus and the minimal tense-logic K_t.

Our new *L*, on the other hand, is not *too* strong for what we require of it, and in particular our proof of *CUabCUbcUab* (transitivity of *U*) in Chapter V, Section 6, will not go through, even with this *L*, if our underlying tense-logic is only K_t, since that proof uses not only *CLpLLp* but *CPPpPp*.

When we pass from K_t to stronger systems, the conditions on *U* which correspond to added tense-logical axioms sometimes involve the function *Iab*, '*a* is the same instant as *b*', which therefore requires some interpretation in the world-calculus if the present methods are to be carried further. A natural translation would be *LEab*, but in fact *LCab* (*Tab*) will do, since from this we can prove *LCba* (and so *LEab*) *ad absurdum* as follows:

C (1) *LCab*
C (2) *NLCba*
K (3) *LCbNa* (2, A2)
K (4) *LCaNb* (3)
K (5) *LCaKbNb* (1, 4)
(6) *NMa* (5),

contradicting A1. For this *I* we have *Iaa* (obviously) and also (inductively) *CIabCφaφb*, where $\phi\alpha$ is any function of α that may be constructed within the system.

As an example of the proofs that are now possible we may take that of the *U*-condition

CUabCUabAAUbcUcbIbc,

which is equivalent to

CUabCUacCNUbcCNUcbIbc,

from its tense-logical counterpart

CKFpFqAAFKpqFKpFqFKqFp.

We have

C	(1) *LCaFb*	(= *Uab*)
C	(2) *LCaFc*	(= *Uac*)
C	(3) *NLCbFc*	(= *NUbc*)
C	(4) *NLCcFb*	(= *NUcb*)
K	(5) *LCaAAFKbcFKbFcFKcFb*	(1, 2, Hyp.)
K	(6) *LCbNFc*	(3, A2)
K	(7) *LCcNFb*	(4, A2)
K	(8) *LNKbFc*	(6)
K	(9) *LNKcFb*	(7)
K	(10) *LGNKbFc*	(8, *CLpLGp*)
K	(11) *LGNKcFb*	(9, *CLpLGp*)
K	(12) *LNFKbFc*	(10)
K	(13) *LNFKcFb*	(11)
K	(14) *CLCaFKbFcNMa*	(12)
K	(15) *CLCaFKcFbNMa*	(13)
K	(16) *NLCaAFKbFcFKcFb*	(14, 15, A1)
K	(17) *LCaNAFKbFcFKcFb*	(16, A2)
K	(18) *LCaFKbc*	(5, 17)
K	(19) *LCaFLCbc*	(18, *CKbcLCbc*)
K	(20) *LCaLCbc*	(19, *CFLpLp*)
K	(21) *MLCbc*	(20, A1)
	(22) *LCbc*	(21, *CMLpLp*).

Summing up, and tidying up: With $L^1\alpha$ for $KK\alpha H\alpha G\alpha$, we define *L* as $\Pi n L^n$; we use *a*, *b*, *c*, as variables standing for those propositions for which we have A1. *Ma*, A2. *ALCapLCaNp*, and A3. Σaa; we define *Uab* as *LCaFb*, *Tap* as *LCap*, and *Iab* as *LCab*. This gives us all we need for moving freely in and out of *U*-calculi from the tense-logics to which they correspond. We can also see more clearly the sense in which the B series is definable in terms of the A series but not vice versa. The tensed *p* can only enter the B-series logic as part of the form *Tap* (which, however, is itself tense-logically definable); the B-series logic has no counterpart of the simple tensed *p*.

4. *The uniqueness of the time-series.* The introduction of forms like L^n, with quantification over the numerals, into the object-language, is in some ways a clumsy device, and it may be worth considering its replacement by the use of infinitely long sentences, with forms like $L^\infty p$ as abridgements. For proofs such as the last one in the last section, a simpler basis still is possible, namely one in which we introduce L (*i.e.* the L used in the world-axioms, and in the definitions of T, U, and I) without defining it at all, as a special primitive (with M as NLN) with the postulates S5+$CLpGp$+$CLpHp$. But however handy this may be as a symbolic simplification, to say that we *must* proceed this way, i.e. that this L is not only *not defined* in a particular calculus but is *undefinable* in tense-logical terms, is a move that would have profound effects both philosophically and formally. It would mean that we cannot reduce the U-calculus (the logic of the B series) wholly to tense-logic (the logic of the A series) after all; and this could be regarded as advantageous or disadvantageous, according to our point of view.

If Lp, asserting that p is true 'all over the diagram', i.e. in all the instantaneous world-states there are, is not tense-logically defined, it is possible to raise the question as to whether there are several distinct time-series, not themselves temporally connected. And *only* if Lp is not tense-logically definable can we raise this question; for to define L tense-logically would be to define it by means of past and future tensings (either straightforward ones or more subtle ones like Kamp's Φ and Ψ) which take their start from *our* 'now' (or as I would prefer to put it, from what *really is* now the case). Only in a U-calculus which stands at least partly on its own feet, with at least a non-tense-logically-defined L among its primitives, can we assert or deny that 'our' time-series stands alone. We can then do it, e.g. by asserting or denying that $Lp = \Pi nL^np$. For to assert this, and still more directly to make the equivalent assertion that $Mp = \Sigma nL^np$, will make the postulate Ma (A1) assert that every term in the field of the relation U is in *some* way temporally connected with the present world. If, on the other hand, L is *defined* as ΠnL^n, and is regarded as only intelligible in some such terms, there is just no alternative to this equivalence.

It may be felt that this very fact is an argument in favour of *not* defining L in this way. For is not the question as to whether 'our' time-series (whatever its structure) is unique, a genuine one? I would urge the following consideration against saying that it is, or at all events against saying it too hurriedly: It is only if we have a more-or-less 'Platonistic' conception of what a time-series is, that we can raise this question. If, as I would contend, it is only by tensed statements that we can give the cash-value of assertions which purport to be about 'time', the question as to whether there are or could be unconnected time-series is a senseless one. We think we can give it a sense because it is as easy to draw unconnected lines and networks as it is to draw connected ones; but these diagrams cannot represent *time*, as they cannot be translated into the basic non-figurative temporal language. If we try so to translate them, we produce contradictions which are a kind of inverse of McTaggart's, like 'Right now there are things going on which stand in no temporal relation to what is going on right now', 'There are things going on which neither *are* going on, nor will be going on, nor have been going on, nor even will have been going on, nor have been going to have been going on—not *anything* like that at all—there really are'. We can only avoid stating this hypothesis in such self-contradictory terms by saying that there timelessly 'are' worlds in which, or instants at which, such-and-such is the case, such-and-such has been the case, such-and-such will be the case ($\Sigma aTap$, $\Sigma aTaPp$, $\Sigma aTaFp$), these worlds or instants being temporally unconnected with *this* one (the present one); this talk of worlds and instants being itself irreducible to talk of what is, has been, will be, will have been, etc.

The question as to the uniqueness of the time-series is thus one of quite a different order from the questions as to whether time is endless or ending, discrete or dense or continuous, circular or non-circular, branching or non-branching, etc. For to raise it as a genuine question is not merely to invite us to consider a non-standard tense-logic, but to suggest that there are truths about time which are not tense-logically expressible. It is not, indeed, to deny outright the existence of an A series, or the possibility and worth of a tense-logic, but it is to deny its primacy, and to relativize it to a B series, a sequence of ordered 'positions' which is tenselessly 'there' (and which may well be only one of a number of such series).

This is a point at which McTaggart seems to me to have been a little too light-hearted. He considers, but rejects, an argument that an A series cannot be essential to time because there may be other B series, though not other A series, than our own. He has no difficulty in disposing of this in the case where the other B series is fictitious.[1] The series of adventures of Don Quixote, 'it is said, does not form part of the A series. I cannot at this moment judge it to be either past, present or future. Yet, it is said, it is certainly a B series. The adventure of the galley-slaves, for example, is earlier than the adventure of the windmills'. The answer is easy; the adventure of the galley-slaves was *not* earlier than the adventure of the windmills

[1] *The Nature of Existence*, ch. 33, §§ 319–21.

because neither of them occurred at all. Certainly it is *said that*, and a credulous reader might *believe that*, the one was earlier than the other, but this means that it is said or believed that the one was past when the other was present or occurring. B-series terms, in short, may be replaced by their A-series definitions within the scope of operators like 'It is said that', 'It is believed that', as well as within more straightforward ones. But McTaggart deals less satisfactorily with the argument that 'there might be *in reality* several real and independent time-series'.[1] He admits that if this were so, 'no present would be *the* present'. But then, he replies, 'no time would be *the* time—it would only be the time of a certain aspect of the universe. It would be a real time-series, but I do not see that the present would be less real than the time'. And again, 'if there were any reason to suppose that there were several distinct B series, there would be no additional difficulty in supposing that there should be a distinct A series for each B series'. Right; but these are A series which are in some way definable 'for' various B series or 'times'; the definition cannot go the other way. Nor can the non-unique 'present' of this hypothesis be the pervasive 'present' of a fundamental tense-logic, for which 'It is (now) the case that p' is equivalent to the plain p, and the plain p to 'It is (now) the case that p'. So it seems to me that anyone who insists that the A series is fundamental must just deny this possibility.

I am sure that these observations have some bearing on the topic of the next section, tense-logic in the theories of relativity; I wish I were clearer as to what that bearing is. In anticipating that section, I feel a bit like someone who, having delivered a Berkeleian attack on the differential calculus, will shortly be nevertheless using it. Point-instants (and even events) seem as mythical to me as matter did to Berkeley; and what I understand of the theory of relativity leaves me about as happy as the calculus left him. Still, it's Science, so in the meantime we can only try (as I shall be trying in the next section) to do our sums right, however obscure their meaning; and wait for Weierstrass.

We may now turn to the *formal* consequences of defining or not defining L as ΠnL^n. If there are questions (genuine or spurious) which this definition prevents us from raising (or prevents us from using the symbol L to raise), there are also theorems which it enables us to prove (this is the other side of the same coin). Obviously, the theorem $ELp\Pi nL^np$ (which the definition turns into a mere notational abridgement of $E\Pi nL^np\Pi nL^np$); but, in consequence, much else besides. In particular, the definition yields the metatheorems that

[1] *The Nature of Existence*, ch. 33, §§ 322–3.

(1) if we subjoin *CGpGGp*, *CKKpHpGpHGp*, and *CKKpHpGpGHp* (i.e. the tense-logical expressions of transitivity and non-branching both ways) to K_t, L^1p (= *KKpHpGp*) becomes equivalent to *Lp*; that (2) if we add the first and just one of the other two, $L^2 = L$; and that (3) if we add *CGpGGp*, *CGpHp*, and *CGpp* (postulates for circularity), even the plain *G* = *L*. To prove (1), for example, we first prove CL^1pL^2p as follows:

C	(1) L^1p	
K	(2) *KKpHpGp*	(1, Df. L^1)
K	(3) *HHp*	(2, *CHpHHp*)
K	(4) *HGp*	(2, *CKKpHpGpHGp*)
K	(5) *KKHpHHpHGp*	(2, 3, 4)
K	(6) *HKKpHpGp*	(5, *CKHpHqHKpq* from K_t)
K	(7) HL^1p	(6, Df. L^1)
K	(8) GL^1p	(analogously)
	(9) L^1L^1p	(1, 7, 8, Df. L^1).

By putting L^np for *p* throughout this proof, we can prove $CL^{n+1}pL^{n+2}p$ for any *n*, and this gives us CL^1pL^np inductively, and so $CL^1p\Pi nL^np$, i.e. *CKKpHpGpLp*, by quantification theory.

This result in turn yields new proofs of *U*-conditions from tense-logical postulates. It was remarked by Lemmon in 1965 that there seems to be no purely tense-logical formula which corresponds exactly to the *U*-condition *AAUabUbaIab*, which he calls strict or strong linearity. There are indeed tense-logical formulae, e.g. *CKKpHpGpHGp* and *CKKpHpGpGHp*, which correspond exactly to non-branching in both directions, i.e. to the pair of conditions:

CUabCUacAAUbcUcbIbc
CUbaCUcaAAUbcUcbIbc.

But, Lemmon pointed out, these conditions and these formulae are compatible with there being several unconnected time-series each of which is separately linear; the categorical *AAUabUbaIab* is not compatible with this, and it is this exclusion of the possibility of distinct time-series which no purely tense-logical formula seems to capture.[1] But we do capture it by the formula proved above, *CKKpHpGpLp*, which says in effect that if *p* is and always has been and always will be true, it is true in all the worlds there are. This intuitively excludes a plurality of time-series, and can be used formally to prove the strict linearity condition on *U*. For we can use it

[1] This point is developed in Cocchiarella's thesis, 'Tense Logic', ch. 3, § 4; see his notes 8, 12, and 13.

to prove an absurdity from the denial of all three disjuncts in $AAUabUbaIab$, thus:

C	(1)	$NLCaFb$	$(= NUab)$
C	(2)	$NLCbFa$	$(= NUba)$
C	(3)	$NLCab$	$(= NIab)$
K	(4)	$LCaNFb$	(1, A2)
K	(5)	$LCbNFa$	(2, A2)
K	(6)	$LCaNb$	(3, A2)
K	(7)	$LCaGNb$	(4)
K	(8)	$LCFaNb$	(5)
K	(9)	$LCHFaHNb$	(8, $CLpLHp$, $CHCpqCHpHq$)
K	(10)	$LCaHNb$	(9, $CpHFp$)
K	(11)	$LCaKKNbHNbGNb$	(6, 10, 7)
K	(12)	$LCaLNb$	(11, $CKKpHpGpLp$)
K	(13)	LMb	(A1, RL)
	(14)	NMa	(12, 13),

which contradicts A1. The truth of Lemmon's contention thus depends on whether $CKKpHpGpLp$ is a purely tense-logical formula or not. If L is not tense-logically defined, it is not, and Lemmon's contention stands; but if it is defined as ΠnL^n, it is, and a plurality of time-series is tense-logically excluded.

We have an analogous result with circular time. Circularity in the sense of the transitivity, symmetry, and reflexiveness of U does not in itself preclude there being a number of distinct circular time-series; for this we need $\vdash Uab$, i.e. '*Every* world is earlier than (and later than) every other', and we could obtain this if we had $CGpLp$, proving it *ad absurdum* thus:

C	(1)	$NLCaFb$	$(= NUab)$
K	(2)	$LCaNFb$	(1, A2)
K	(3)	$LCaGNb$	(2)
K	(4)	$LCaLNb$	(3, $CGpLp$)
K	(5)	LMb	(A1, RL)
	(6)	NMa	(4, 5),

contradicting A1. And we do obtain $CGpLp$ from the usual circularity axioms if we define L as ΠnL^n, but if we take it as undefined (with postulates $S5+CLpHp+CLpGp$), we do not.

In a U-calculus which is not tense-logically anchored, but is taken as basic, and in which tensed formulae (and formulae with L) occur only as second arguments of the functor T (i.e. in which the tensed *proposition* α is replaced by the tensed *predicate* $T'\alpha$), some of the last-mentioned results may be presented as follows: To the usual basic equivalences U1–U4 we add

$$U5: ETaLp\Pi bTbp.$$

It is well known that with this for *L* we have all S5 theses preceded by *Ta*; and it is easy to prove *TaCLpGp* and *TaCLpHp*. The converses *TaCGpLp* and *TaCHpLp* are provable if we add $\vdash Uab$ to our basis, i.e. if we equate *U* with the universal relation; and *TaCKKpGpHpLp* if we lay down $\vdash AAUabUbaIab$, i.e. if we equate $(U \cup \breve{U} \cup I)$, the logical sum of *U*, its converse and identity, with the universal relation. Unconnected *U*-series would be excluded by laying it down that the ancestral of the logical sum of *U* and its converse and identity relates every *a* and *b*, i.e. $\vdash(U \cup \breve{U} \cup I)_* ab$, or $\vdash(U \cup \breve{U} \cup I)_* \doteq \dot{V}$. Given this, we may prove $\vdash Uab$ from *U*'s being transitive, reflexive, and symmetrical, and $\vdash AAUabUbaIba$ from its being transitive and non-divergent both ways.

5. *The tense-logical discrimination of special from general relativity.*[1] It has now become almost a commonplace that if we use $L\alpha$, following Diodorus, for $K\alpha L\alpha$, then (*a*) if our tense-logic is geared to the earlier-later relation of classical physics, the resulting Diodorean-modal system is S4.3, whereas (*b*) if our tense-logic is geared to the earlier-later relation, or one of the earlier-later relations, of relativistic physics, the resulting Diodorean-modal system is S4. This seems to me to need a small correction, and I would suggest that while S4 does indeed give the Diodorean-modal logic appropriate to the *general* theory of relativity, the Diodorean-modal logic appropriate to the *special* theory is at least S4.2.

The position appears to be as follows. Both theories of relativity admit a 'local proper time' which is linear, and so yields a tense-logic with S4.3 as its Diodorean-modal fragment, but there is not just one but an indefinite number of such 'local proper times', and a distant event *b* may be earlier than an event *a* in the frame of reference associated with one such 'proper time', and later in another. This, however, is only true within limits, and in some cases an event *b* is earlier or later than an event *a* with respect to *all* frames of reference, and so may be said to be 'absolutely' earlier or later. In particular, if the space-time points *a* and *b* could conceivably be linked by the path of a light-signal, one of them will be absolutely earlier than the other, and the other absolutely later. It is for this public or causal relativistic time that we can construct tense-logics with the other Diodorean-modal fragments mentioned.

[1] In this section I am indebted to Mr. E. E. Dawson for checking my physics.

The difference between S4, S4.2, and S4.3, it will be recalled, is that the weakest system S4 assumes only that the earlier-later relation is transitive, the strongest system S4.3 in addition precludes branching, and the intermediate system S4.2 does not completely preclude branching, but does preclude it unless the branches eventually meet again. It is not immediately obvious that the line-patterns associated with these theories have anything to do with the theory of relativity; but forget this picture; we are now on to another, in which we are concerned not with the meeting of lines but with the eventual overlapping of ever-enlarging illuminated volumes. In terms of *U*-calculi—of pure algebra, as it were—the condition corresponding to the S4.2 axiom *CMLpLMp* (the underlying tense-logical axiom would be *CFGpGFp*) is given by

$$CUabCUac\Sigma dKUbdUcd,$$

'If both *b* and *c* are in *a*'s future, then there is some *d* which is in the future of both of them', or 'If *b* and *c* are both later than *a*, then some *d* is later than both of them'. If we read *Uab* as asserting that the space-time point *b* is within the forward 'light-cone' of *a*, the above formula will assert that if two space-time points *b* and *c* are both within the forward light-cone of some point *a*, then there is some point *d* which is within the forward light-cone of both of them. The point is simply that all the forward light-cones eventually intersect one another. We have a pattern more or less like this:

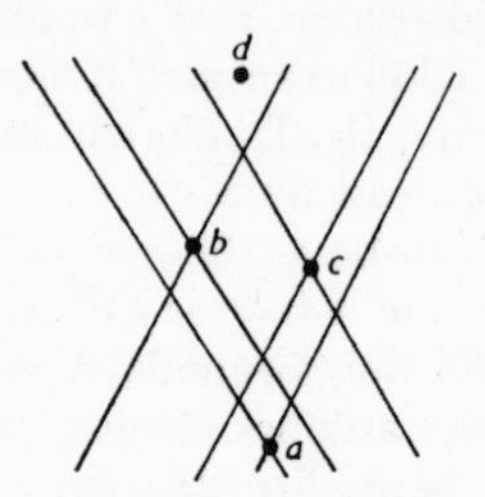

Here the points *b* and *c* are both within the forward light-cone of *a*, and however distant they are, their own forward light-cones will eventually intersect, and there will be points such as *d* within both, which will therefore be absolutely future to both of them. And if, at *a*, it will be the case, say at *b*, that something or other will always be the case (will fill all of *b*'s forward light-cone), then at *a* it will always be the case, i.e. it will be the case at any point *c* within *a*'s forward light cone, that the thing will eventually be the case somewhere in *c*'s forward light-cone, namely after *c*'s cone enters *b*'s (e.g. at *d*); or in short, *CFGpGFp*. This condition is met in the space-time of special relativity; at least it is met if we assume that time has no end. (In this space-time, we might say, all futures tend to merge, but if time stopped some futures would be left separated.) In general relativity, however, the condition need not be met, as that theory allows for the possibility of light 'cones' which so twist away from one another that after a while they never intersect at all.

Special-relativity tense-logic would seem to be obtained by adding to K_t the axioms *CGpGGp*, *CGpFp*, *CFGpGFp*, and the mirror images of the latter two; with *CGGpGp* if we wish to assert denseness.

In both relativistic theories there are points or 'worlds' which are neither in the past nor in the future of a given point or 'world', nor yet identical with it, though they will be connected with it by some sequence of *P*s and *F*s—in our diagram, for example, it will be true at *b* that it will be the case (at *d*, for instance) that it has been the case that *c*, and also that it has been the case (at *a*, for instance) that it will be the case that *c*; and vice versa. In special relativity, in fact, we have the theorem that whatever is the case anywhere in space-time will have been the case—in the notation of the last section, $C\Sigma mM^{m}pFPp$, or *CMpFPp*; with its mirror image. From *CMpFPp* and ΠaMa (A1 of world-logic) we easily obtain $\Pi aFPa$, which is very like Findlay's own symbolizing of his theorem *CAApPpFpFPp*; the Findlay formulation is in fact correct for worlds, both classically and in special relativity. We could also say, in special-relativity tense-logic: $CMpM^{2}p$; or $CL^{2}pLp$. In *dense* special-relativity tense-logic, we even preserve Hamblin's 15-tense theorem.

The key formula $CUabCUac\Sigma dKUbdUcd$ may be proved from *CFGpGFp*, using the methods of the last two sections, as follows: ⊢*CFGpGFp* yields ⊢*LCFGpGFp* by RL, and this yields ⊢*CTaFGpTaGFp* by *CLCqrCLCpqLCpr* and $T = LC$. This in turn gives us, by U1–U4,

$$C\Sigma bKUab\Pi eCUbeTep\Pi cCUac\Sigma dKUcdTdp$$

which is deductively equivalent, by quantification theory, to

$$CUabC\Pi eCUbeTepCUac\Sigma dKUcdTdp.$$

Substitution of *Pb* for *p* in this, and the definition of $U\alpha\beta$ as $T\beta P\alpha$, yields

$$CUabC\Pi eCUbeUbeCUac\Sigma dKUcdUbd,$$

from which the second antecedent $\Pi eCUbeUbe$ may be detached, giving (apart from a permutation of conjuncts at the end) the required formula.

6. *Alternative axioms for non-branching.* I shall show in this section the deductive equivalence, with respect to K_t, of the following three formulae

A. *CKFpFqAAFKpqFKpFqFKqFp*
B. *AGCpCGpqGCGqp* (due to C. Howard)
C. *CPFpAApPpFp*

I shall prove A from B, and B from A and C. (Lemmon's proof of C from A is given in Chapter III.)

In proving A from B, we first prove the following modification of it:

D. *CGCpqCGCpGqCGCNqGNpCNGNpGq*,

i.e. we prove that *GCpq*, *GCpGq*, *GCNqGNp*, and *NGNp* jointly imply *Gq*. By B, these four antecedents, if they are all true, must either be true together with *GCqCGqNp* or together with *GCGNpq* (since substitution in B gives *AGCqCGqNpGCGNpq* as a law). But not the former, since *GCpq* and *GCqCGqNp* yield (by *GC*-syllogism) *GCpCGqNp*, and so *GCGqCpNp*, and so *GCGqNp*. But this, with *GCpGq*, yields *GCpNp*, and so *GNp*, contradicting the last antecedent *NGNp*. So they can only be true in conjunction with the other alternative *GCGNpq*. But this, with the antecedent *GCNqGNp*, yields *GCNqq*, and so *Gq*, the final consequent. From D, now proved, we obtain A by elementary transpositions thus:

D = *CNCNGNpGqNKKGCpqGCpGqGCNqGNp*
= *CNCNGNpGqAANGCpqNGCpGqNGCNqGNp*
= *CKNGNpNGqAAFKpNqFKpNGqFKNqNGNp*
= *CKFpFNqAAFKpNqFKpFNqFKNqFp*,

which yields A by the substitution *q/Nq* and double negation.

In proving B from A we first transform it by elementary transpositions into

NKFKpKGpNqFKGqNp,

and prove that the conjunction here denied is impossible. For by A this conjunction, i.e. *K(FKpKGpNq)(FKGqNp)*, entails

AA (1) *FK(KpKGpNq)(KGqNp)*
(2) *FK(KpKGpNq)(FKGqNp)*
(3) *FK(FKpKGpNq)(KGqNp)*.

Here the alternative (1) is impossible because it asserts the future truth of a conjunction in which one component is *p* and another *Np*; (2) is impossible because the second main conjunct entails *FNp*, which contradicts *Gp* in the other main conjunct; and (3) because the first main conjunct entails *FNq*, contradicting *Gq* in the other.

In proving B from C we again do it by proving the impossibility of the conjunction *K(FKKpGpNq)(FKGqNp)*. By *CpGPp* the second conjunct entails *GPFKGqNp*, and by *CKFpGqFKpq* this with the first conjunct yields

$$FK(KKpGpNq)(PFKGqNp),$$

and this by *C* yields

$$FK(KKpGpNq)(AAKGqNpPKGqNpFKGqNp),$$

and so

$$FAA\ (1)\ KKKpGpNqKGqNp$$
$$(2)\ KKKpGpNqPKGqNp$$
$$(3)\ KKKpGpNqFKGqNp.$$

Here the alternative (1) is never possible, since *p* and *Np* are both among its conjuncts; (2) because *PKGqNp* entails *PGq* and so *q*, contradicting the conjunct *Nq*; and (3) because *FKGqNp* entails *FNp*, contradicting the conjunct *Gp*.

These proofs make it clear that C may replace A not only in a comparatively strong tense-logic such as Scott's, but also with no auxiliary assumptions but those of K_t.

Non-branching in both directions is given not only by the combination of one of these axioms with its mirror image but also by laying down the S4 law *CMMpMp* for $M\alpha$ defined as $AA\alpha P\alpha F\alpha$. (If this works, the stronger S5 law *CMNMpNMp*, of which it was noted in 'The Syntax of Time-Distinctions' that it seems to assume non-branching, will clearly work also. We just prove *CMMpMp* from it in the usual way, and then proceed as below.) By Df. *M*, *CMMpMp* expands to

$$CAA\ \left.\begin{array}{l}(1)\ AApPpFp\ \ (=Mp)\\(2)\ PAApPpFp\ \ (=PMp)\\(3)\ FAApPpFp\ \ (=FMp)\end{array}\right\}(=MMp)$$
$$(4)\ AApPpFp\ \ (=Mp).$$

This is deductively equivalent to the three theses *C* (1) (4), *C* (2) (4), *C* (3) (4), of which the first may be dropped, being a mere substitution in *Cpp*. Then *C* (2) (4) = *CPAApPpFpAApPpFp* = *CAAPpPPpPFpAApPpFp* → *CPFpAApPpFp*, and the mirror image is proved from *C* (3) (4) similarly.

7. *Tenses defined in terms of Diodorean modalities.* In Chapter V, Section 5, it was shown that if we use an *L* for which we have at least the system T, and define *Gp* as *ΠqCqLCNqp*, we can prove at least the postulates of the future-tense portion of K_t (i.e. the postulates of Lemmon's modal system T(C) with *G* for *L*), and can also prove the equivalence of *Lp* to *KpGp* (cf. Diodorus). The question was then raised as to the deducibility of stronger tense-logics from correspondingly stronger modal logics, given this definition. For instance, if we add *CGpGGp* to K_t, and define *Lp* as *KpGp*, the resulting modal logic is known to be S4, i.e. T+*CLpLLp*; if, conversely, we start from S4, and use the above definition of *G*, do we obtain K_t+*CGpGGp*? This particular question, at least, can now be answered in the negative.

We may note to begin with that if *Lp* is equivalent to *KpGp*, *CLpLLp* will be equivalent to *CKpGpGGp* (*CLpLLp* = *CKpGpKLpGLp* = *CKpGpKKpGpGKpGp* = *CKpGpGKpGp* = *CKpGpKGpGGp* = *CKpGpGGp*). So part of our problem is: Given K_t, could *CKpGpGGp*, or *CpCGpGGp*, be laid down as a thesis without *CGpGGp* becoming one? The answer is that it certainly could if we had *CpGGp*; for since *KpGp* implies both *p* and *Gp*, it will imply *GGp* if either of those does. One way of obtaining a system with *CpGGp* (and so *CpCGpGGp*) but not *CGpGGp* is to suppose there are only two world-states, and let *Gp* be true in a given state if and only if *p* is true in the other one. We would then have *CpGGp*, which would now assert that what is true in a given state is true in the other of the other one, i.e. in the given one; but we would not have *CGpGGp*, which would assert that what is true in the other state is thereby true in the other of the other, i.e. in the given one.

We may give this independence proof a more formal character by using the following Meredith-style 4-valued matrix in which the value 1 means 'true in both worlds'; *n* means 'true in world *n* only'; $\bar{n}$, 'true in $\bar{n}$ only'; and 0, 'true in neither':

C	1	*n*	$\bar{n}$	0	*N*	*G*	*L*
* 1	1	*n*	$\bar{n}$	0	0	1	1
n	1	1	$\bar{n}$	$\bar{n}$	$\bar{n}$	$\bar{n}$	0
$\bar{n}$	1	*n*	1	*n*	*n*	*n*	0
0	1	1	1	1	1	0	0

It will be found that the column for *Lp* is what we would get by defining it as *KpGp* and using the column for *G*, and that *CLpLLp*, but not *CGpGGp*, = 1 for all values of *p*. This matrix exactly characterizes a 'tense-logic' defined by the following axioms (subjoined to propositional calculus with substitution, detachment, and RG):

A1. *CGCpqCGpGq* A2. *CNGpGNp*
A3. *CGNpNGp* A4. *CpGGp*.

A1–3 are what one gets by putting *G* for Scott's monadic *T*; A4 expresses the special character of this *G*, and its converse is easily obtained from it by A2 and 3. (A4 gives *CNpGGNp* by substitution, and this $= CNGGNpp = CGNNGpp = CGGpp$.) Any verification of a formula by the 4-valued matrix may be turned very simply into a deduction from these postulates. For in the first place, the four possible assignments of values can be expressed within the system as follows (taking n to be the present world and $\bar{n}$ the other):

$p = 1$ (i.e. p true in both) as *KpGp*
$p = n$ (i.e. p true in n only) as *KpNGp*
$p = \bar{n}$ (i.e. p true in $\bar{n}$ only) as *KNpGp*
$p = 0$ (i.e. p true in neither) as *KNpNGp*.

The basic evaluations summed up in the matrix may then be expressed as provable implications, as in the following samples:

$Gn = \bar{n}$ means: If $p = n$, $Gp = \bar{n}$, i.e. if *KpNGp* then $KN(Gp)G(Gp)$: *CKpNGpKNGpGGp*, provable from p.c. and A4 (*CpGGp*).
$Nn = \bar{n}$ means: If $p = n$, $Np = \bar{n}$, i.e. if *KpNGp* then $KN(Np)G(Np)$: *CKpNGpKNNpGNp*, provable from p.c. and A2 (*CNGpGNp*).
$C\bar{n}0 = n$ means: If $p = \bar{n}$ and $q = 0$ then $Cpq = n$, i.e. if *KNpGp* and *KNqNGq* then *KCpqNGCpq*: *CKNpGpCKNqNGqKCpqNGCpq*, provable from p.c. and A1 transposed to *CGpCNGqNGCpq*.

The use of the matrix to evaluate more complex formulae, e.g. the calculation $CGnGGn = CGnG\bar{n} = C\bar{n}n = n$, can be mirrored by deductions from the implications enshrined in the matrix; in this case we prove that if $p = n$, $Gp = \bar{n}$, and *GGp* consequently n, and *CGpGGp* consequently n, i.e.

C (1) *KpNGp* $(p = n)$
K (2) *KNGpGGp* $(Gp = \bar{n})$
K (3) *KGGpNGGGp* ($GGp = n$; from 2 by $G\bar{n} = n$, i.e. *CKNpGpKGpNGGp*, with *Gp* put for p)
(4) *KCGpGGpNGCGpGGp* ($CGpGGp = n$; from 2 and 3 by $C\bar{n}n = n$, i.e. *CKNpGpCKqNGq*—*KCpqNGCpq*, with *Gp* for p and *GGp* for q).

Finally, if $f(p)$ works out as 1 for all values of p, this means that if $p = 1$, n, $\bar{n}$ or 0 then $f(p) = 1$, i.e. if *KpGp* or *KpNGp* or *KNpGp* or *KNpNGp* then $Kf(p)Gf(p)$; this being proved (disjunct by disjunct) by the above methods, we get $Kf(p)G(p)$ unconditionally, and so $f(p)$, by detaching *AAAKpGpKpNGpKNpGpKNpNGp*, which is a substitution in a p.c. theorem. The extension of this procedure to cases involving more than one variable is fairly obvious.

This, however, is a *G*-primitive system, so that we have not yet *quite* shown that we can have *CLpLLp* without *CGpGGp* if we take *L* as primitive and define *Gp* as *ΠqCqLCNqp*. We may justify this last step by observing that when 1, *n*, *n̄* and 0 are all the *q*'s (or 'values of *q*') that there are, *ΠqCqLCNqp* amounts to

$$KKK(C1LCN1p)(CnLCNnp)(C\bar{n}LCN\bar{n}p)(C0LCN0p)$$

which with the given column for *L* works out as 1, *n̄*, *n*, 0 when $p = 1, n, \bar{n}, 0$ respectively, exactly as in the given column for *G*.

It was obvious all along that added *G*-theses cannot always be got back from resulting added *L*-theses, since in some cases the latter do not exist; e.g. if we add *CGGpGp* (density) or *CGpFp* (non-ending) to K_t, this does not enrich the Diodorean *L* system in any way (we already have *CLLpLp* and *CLpMp* in T, the Diodorean fragment of K_t). What is now clear is that even when the strengthening of the tense-logic does strengthen its Diodorean-modal fragment, we do not, or at all events do not always, get the tense-logical strengthening back when we start from the resultant strengthening of the modal system.

This does not mean that we cannot have *L*-primitive, *G*-defined tense-logics containing such theses as *CGpGGp*, *CGGpGp* and *CGpFp*. We can obviously obtain such systems simply by laying down as axioms the definitional expansions of these theses, e.g. by laying down *CGpGGp* in the form

$$C\Pi qCqLCNqp\Pi rCrLCNr\Pi sCsLCNsp.$$

But we cannot obtain them by laying down *L*-theses (valid in the tense-logics concerned) which do not contain propositional quantifiers. Short of that, however, we can sometimes make instructive simplifications. For example, the formula *FCpp* is deductively equivalent (given K_t) to *CGpFp*, and may replace it as an expression of non-endingness. Geach's definition of *F* in terms of the Diodorean *M* (equivalent to the above definition of *G* in terms of *L*) turns *FCpp* into *ΣqKqMKNqCpp*. But since *KrCpp* is interchangeable (even in T) with the plain *r*, this may be simplified to *ΣqKqMNq*, 'For some *q*, it is the case that *q*, but (is or) will be the case that not *q*'. Considered as a version of 'There is more time to come' (as entailed by this as well as entailing it), this very nicely reflects McTaggart's 'There could be no time if nothing changes', the original inspiration of Geach's definition. Again, in the above expansion of *CGpGGp*, only the first quantifier is essential, given S4 for *L*.

8. *Independence proofs for* K_t. Hacking and Berg have the following independence proof for *CGCpqCGpCq*: Let *k* be some true proposition which is 'atomic' with respect to the functions of the system (i.e.

it is not a negation or a tensing of one of the propositions of the system, or an implication of one of them by another) and is not an axiom or theorem of the system. Let us write '$p = q$' for 'p and q have the same truth-value' and 'p is q' for 'The proposition that p is the same proposition as the proposition that q'. Let $Hp = p$ for all p, and let $Gp = p$ except when p is k, and then let $Gp = 0$. (Gp thus amounts to 'It is the case that p, and p is not k'.) Since k is atomic, Np is never k, so GNp always $= Np$, and $Fp = NGNp = NNp = p$. Since k is not a theorem, we never have $\vdash k$, so $\vdash \alpha$ always gives $\vdash G\alpha$. $CHCpqCHpHq = CCpqCpq = 1$; $CFHpp = Cpp = 1$; $CPGpp = Cpp$ (when p is not k) or Cop (when p is k), which in both cases $= 1$. But when q is k, and $p = 1$ but is not k, $CGCpqCGpGq = CGCpkCGpGk = CCpkCpo = CC11C10 = C10 = 0$. This interpretation also verifies

$$CGpGGp, \quad CGGpGp, \quad CGpFp, \quad CKFpFqAAFKpqFKpFqFKqFp$$

and their mirror images, so that $CGCpqCGpGq$ is independent of these also. Independence of $CHCpqCHpHq$ may be established by interchanging the roles of G and H. If we axiomatize with a mirror-image rule instead of mirror images of the axioms, we can use the same model but with $Hp = Gp$ instead of $Hp = p$.

Where the system is axiomatized with a mirror-image rule, $CPGpp$ can be proved independent by letting $H = G$ and so $P = F$, and otherwise interpreting the symbols normally (Hacking and Berg). This turns $CPGpp$ into $CFGpp$, which is not a law of normal future-tense logic. (It is obvious that p may be going-to-be-always-true without being true now.) If we have no mirror-image rule, but lay down $CPGpp$ (or $CpHFp$) and $CFHpp$ (or $CpGPp$) separately, we can prove them separately independent by using the following modification of the U-calculus (due to Lemmon, 1965): Let us have worlds or instants ordered not by one but by two relations, say U and Y, and let

$$TaGp = \Pi bCUabTbp$$
$$TaHp = \Pi bCYabTbp.$$

If we read Uab as 'b is later than a' and Yab as 'b is earlier than a', these amount to:

> Gp is true at a if and only if p is true at all instants later than a
> Hp is true at a if and only if p is true at all instants earlier than a.

Normally, of course, we suppose that Y is simply the converse of U, i.e. $Yab = Uba$, but let us drop this assumption, and replace it by he one-way implication $CYabUba$. $TaCpHFp$, i.e.

$$CTap\Pi bCYab\Sigma cKUbcTcp,$$

is then provable thus:

ΠbC (1) *Tap*
C (2) *Yab*
K (3) *Uba* (2, *CYabUba*)
K (4) *KUbaTap* (1, 3)
(5) *ΣcKUbcTcp* (4, E.I.).

But in the absence of *CUabYba*, a similar proof of *TaCpGPp* is impossible. If, conversely, we lay down *CUabYba* but not *CYabUba*, we can prove *TaCpGPp* but not *TaCpHFp*. The provability of the *U*-theses corresponding to the other postulates of K_t is obviously unaffected by this modification.

9. *Anticipations of later developments in Łoś's calculus of instants.* In Chapter I, in listing the precursors of modern tense-logic, I ought not to have omitted the calculus which Jerzy Łoś devised in 1947 in an attempt to formalize Mill's canons of induction. The calculus appeared in the *Annales Universitatis Mariae Curie-Sklodowska*, Section F, vol. 2 (for 1947, published in 1948), pp. 269–301, and was summarized and reviewed by Henry Hiż in the *Journal of Symbolic Logic*, vol. 16, No. 1 (March 1951), pp. 58–59. (I only know the paper through Hiż's review.) Łoś's calculus has no tense-operators, but does use propositional variables p_1, p_2, etc., to stand for what might be 'satisfied' at one instant and not at another. He also has variables t_1, t_2, etc., to stand for instants and n_1, n_2, etc., for temporal intervals; the form Ut_1p_1 for 'p_1 is satisfied at t_1', and $\delta t_1 n_1$ for 'the instant n_1 later than t_1'. He abridges $\Pi p_1 EUt_1p_1Ut_2p_1$ to $\rho t_1 t_2$, which may be read as 't_1 and t_2 are the same instant'. This calculus influenced my own formulation of a 'calculus of dates' (using the form *Utp*) in *Time and Modality*, and also has points of resemblance to Rescher's systems of 1965. It will facilitate comparisons if we give Łoś's axioms in the symbolism of Chapter 6, Section 4, supplemented by *San* for 'the instant *n* later than *a*', and *Iab* for '*a* and *b* are the same instant'. (*Iab*, it should be remembered, is short for *ΠpETapTbp*, and *TSanp* is equivalent to *TaFnp* in the symbolism of Chapter 6, Section 4.) The axioms then become:

1. *ETaNpNTap*
2. *CTaCpqCTapTaq*

3, 4, and 5. *TaCCpqCCqrCpr*, *TaCpCNpq*, *TaCCNppp*
6. *CΠaTapp*
7 and 8. *ΣbISanb*, *ΣbISbna*
9. *ΣpΠbETbpIab*.

Axioms 1–5 have as consequences all propositional-calculus theorems preceded by *Ta* (cf. the rule, in Rescher's systems and mine, to infer $\vdash Ta\alpha$ from $\vdash\alpha$). 7 would seem to be replaceable by the permission to substitute any expression of the form *San* for instant-variables in theses, though it brings out the fact that this permission assumes that there is an instant at any arbitrary interval after, as 8 asserts that there is at any interval before, any given instant. Łoś apparently thought that this requires 'that there be an infinite number of constants which can be substituted for the variables representing instants'; Hiż argued in his review that this would only be the case if we had an axiom, say *CISanbNIab*, excluding circularity. Axiom 9, the 'clock axiom', asserts in effect that 'to every instant of time a function can be assigned (e.g. the description of the position of the hands of a clock) which is satisfied only at that instant'. Łoś regarded it as 'our only weapon against the metaphysical and extrasensual conception of time'. His point of view seems in fact to have been very close to that of Sections 3 and 4 of this appendix. The clock axiom, one might say, might justify (or might reflect) our *identification* of an 'instant' with a proposition true at that instant only; it corresponds to the postulates *Taa* and *CTabIab* in a system using instant-variables as a special sub-class of propositional ones.

Łoś found that his axiom rather trivialized his formulation of Mill's canons, and thought they might appear less trivial as consequences of an 'axiom of causality' which he formulated as *ΣpETSanpTaq*, asserting that for any *a*, *q*, and *n*, there is a *p* which is true *n* later than *a* if and only if *q* is true at *a*. But given tenses, this is trivial also, since *Pnq* will automatically meet this condition. So will *Taq*, given nested *T*-ing with the normal law *ETbTaqTaq* (true at *b* that *q* is true at *a*, if and only if *q* true at *a*).

INDEX

PRINTED IN GREAT BRITAIN
AT THE UNIVERSITY PRESS, OXFORD
BY VIVIAN RIDLER
PRINTER TO THE UNIVERSITY